THE GOLDEN TOAD

AN ECOLOGICAL MYSTERY AND THE SEARCH FOR A LOST SPECIES

TREVOR RITLAND & KYLE RITLAND

DIVERSION BOOKS

Diversion Books
A division of Diversion Publishing Corp.
www.diversionbooks.com

Copyright © 2025 by Kyle and Trevor Ritland

All rights reserved, including the right to reproduce this book or portions thereof in any form whatsoever. No part of this publication may be reproduced or transmitted in any form or by any means, electronic or mechanical, including photocopying, recording, or any other information storage and retrieval, without the written permission of the publisher. Diversion Books and colophon are registered trademarks of Diversion Publishing Corp.

For more information, email info@diversionbooks.com.

Hardcover ISBN: 978-1-63576-996-8
e-ISBN: 978-1-63576-991-3

First Diversion Books Edition: June 2025
Cover design by Jonathan Sainsbury / 6x9 design
Design by Neuwirth & Associates, Inc.
Chapter illustrations by Daniel Wesson

Printed in the United States of America
1 3 5 7 9 10 8 6 4 2

Diversion books are available at special discounts for bulk purchases in the US by corporations, institutions, and other organizations. For more information, please contact admin@diversionbooks.com.

The publisher does not have any control over and does not assume any responsibility for author or third-party websites or their content.

For LBRP and BB

. . .

CONTENTS

Arcadia	vii
Prologue: An Orange Bufo *Forest*	xi
I. Ghost Stories	1
II. The Golden Age and the Green Mountain	19
III. The End of the Show	49
IV. The Creeping Fear	77
V. In the Forests of the Lost Frogs	101
VI. Into Brillante	121
VII. The Survivors	143
VIII. On the Mountain of Revelation	171
IX. Burial of the Dead	197
X. The Eternal Forest	221
Epilogue: El Dorado y Los Ángeles	247
Acknowledgments	255
Endnotes	258

ARCADIA

It was swirling mist, descending from the heavens under cloud drifts, that gave the mountain life. In the distant ranges, a great lake glowed red at the foot of a volcano, blue night broken by an ancient fire. The lower cordilleras were dry, running down in wide banks to meet the sea, but in the higher elevations the mist and fog washed down across the wind-worn forests, carrying the precious jewel of water. On the ridgeline of those highlands, small streams wound down through shallow canyons and wore tracks in the face of the old mountain over eons. Trees collapsed under the weight of climbing plants and decomposed, returning to the soil. Creeping fungus and bright mushrooms washed over the understory like a slow tide, advancing and retreating, finding damp pockets of shadow. Small creatures were born and died and decayed, wet green mosses covering their tombs. The understory shimmered with soft light refracted in kaleidoscopes of fractals by beads of dew. The drifting mist fell on the leaves of creeping ferns and grasping vines, and into bright flowers high in the canopy of the early trees, running down the woody bodies and living stems to reach the soft, sweet earth. Among the roots, in hollows of dirt and plant and rock, small pools collected, just a few fingers deep, and shimmered with the silver light of the distant moon.

In one of those pools, a clutch of eggs was tucked into the cradle of the firmament. Round and smooth, dark like the earth, they lay in a small cluster, reflecting light in ribbons as it glanced through the leaves and branches that hung above. Undisturbed and unnoticed, their forerunner long vanished, the clutch relied on chance and luck and the law of the jungle in the mystic cloud forest. If the evening rain had come down stronger, they would have been washed away into the lowlands, never to be seen again; if the early-morning light had reached the understory, their pool would have evaporated, and they would have been lost. A clutch of dark eggs drying, or separated and scattered—life blinking out—would not have altered the earth's slow spin; but there, on that mountain at that moment, the elements had conspired for their survival. And sometime in one of those long dark and quiet midnights, cold air blowing up the mountain, the first golden toad emerged.

The first of them would have been alone. Knowing nothing like itself, it had diverged from a winding lineage, reaching back into the muddy banks of evolution, infused with an ancestral memory that soothed its aching lonesomeness in the forests of solitude. It grew from a tadpole in its shallow pool, tasting life in an inch of water; too small to be observed, it evaded roaming predators long enough to transform. In the dripping dark, it turned golden—painting the understory like a spark of fire. There it waited, not knowing for what; it could only glimpse the deep black sky beyond wreaths of foliage, watch small lights shining, and wish to no longer be alone. In time, brothers and sisters emerged, chirping their creation myths. The remnants of the clutch of eggs decayed into the earth, and the golden toads made their home on the high mountain.

In the same epoch, *Homo sapiens* ventured into the dark reaches of the continent, cutting a winding path from a long-forgotten homeland. The golden toads could only guess what gifts or curses this new species might carry through the forest. As strange voices echoed in the canyons—our fortune-telling, our ghost stories—they would have watched their generations live and die as they handed down their precious secrets, teaching the young tadpoles all their knowledge of the wild forest. They could never have known that the god of death was already near, climbing slowly and softly up the mountain, bag on its back and greed burning in its eyes. In the long death waltz, its path would lead it deep into the hearts of the golden toads, and the mountaintop would glow no more with their illumination. Then the golden toad would become a ghost, glimpsed in memory or as a phantasm of arcadia, and known only to a few who had witnessed its ethereal existence in the mystic forest high among the clouds. That story would be passed down in pieces, in fractured impressions and faded memories: lost tales told over a dying fire.

Now the wind blows up the mountain like a banshee, and the crooked arms of wood drip water on a forest floor once abounding with the golden totems, tucked away like myths among the leaves and roots. And from time to time the old forest stirs with the memory: heavy mist creeping among contorted trees, green moss beds glinting with the dew of eternity, a thunder of rain over distant mountains, and an orange glow high on the continental ridge as the golden toads emerged.

PROLOGUE

AN ORANGE BUFO FOREST

San José, Costa Rica

The runway of the new El Coco Airport in Costa Rica's Central Valley was a white scar of cement over the green earth, sprawling toward the end of a stony field where big four-engine planes brooded dreamlessly in dark hangars. On the slow descent above the Pass of Desengaño, Jay Savage looked down through ghostly clouds to see San José nestled in the valley between the central cordillera* and the low rise of the Talamanca Mountains, which wound southeast toward Panama. Old stone churches near the capital's heart gave way to sheet-metal houses in the soup of the outer city. That year, Irazú was erupting—spewing ash over the countryside—and the pilots watched the smoldering volcano distrustfully as they brought the airplane down, silver steam drifting from the caldera's open mouth.

It was April 1964, and Savage was returning to Costa Rica after only six months away; he had left the country at the start of the windy-misty season, but as the plane descended to carry him back into the tropics, the

---

* Cordillera: A mountain range or a system of mountain ranges.

proper wet season was washing over the mountains like night surf. The first real storms would be arriving soon, swallowing the dark forests in water and filling the rivers with mud-like coffee. The rain would call the most secretive frogs out from their hidden dens high on the mountain slopes, and the forest would transform into a dim, wet labyrinth wrapped in creeping fog. As the plane touched down on Costa Rican soil, Savage hoped that he could reach his destination before the rains stirred mudslides on the mountain roads and closed the pass for the season.

After talking his way through customs, strange chemicals and equipment rattling on his back, he moved through the city with the efficiency of purpose. It was already late in the afternoon, and this close to the equator it would be dark before long. The roads were overrun with evening traffic, but he was able to hail a taxi that would take him to the University of Costa Rica's campus on the outskirts of San José, where a golden treasure was waiting to be examined.

For the last few years, Savage had been entangled in an informal competition with a Kansas herpetologist named Bill Duellman to document the known and unknown frogs and toads of Costa Rica's rainforests. This new revelation, if true, would bring that competition to a close in no uncertain terms. A few months earlier, Savage had heard a story from a young man at the Quaker colony up in the mountains of Monteverde, northwest of San José; Jerry James had spoken of a new toad unknown outside the small community in the high mountains, bright orange and visible in the hundreds through the murky tangles of a gnarled forest bent by the eternal wind.

At first, Savage had been skeptical of the report. The toads of the Americas were known to be generally unremarkable in appearance; and though many tropical forests were still undocumented by North American biologists, collectors had traveled relatively deep into Costa Rica by the 1960s. The aggregation numbers that James was describing—hundreds, if not thousands, of the colorful toads—were unlike anything that had ever been recorded. Savage had his doubts

PROLOGUE

about the story and the species; it wouldn't be the first time that a local's fantastic claims turned out to be little more than legend. Norman Scott, a graduate student of Savage's who was working at the University of Costa Rica at the time, put it more bluntly: "We didn't believe him."

As his taxi navigated the city, the smell of diesel and grilled meat washed along the streets. Jay Savage flipped through his field notes from the previous October, when Jerry James had led him up to the town called Monteverde, only to find no sign of the toads. James had explained that the season was wrong, that they only emerged for a few weeks every year; where the toads went when they were not gathering in the hidden breeding grounds, no one seemed to know. Ever since, the story of the golden toad on the green mountain had been on Savage's mind, and the cloud forests* certainly had their secrets. The possibility was entrancing—and as far as he knew, Duellman had not yet gotten wind of it—so Savage had returned to Costa Rica with the early rains for another field season, the mysterious species lingering in his thoughts.

It was evening by the time Savage reached the UCR campus, where he was meeting Norman Scott to plan the specifics of their next collecting trip. Scott had taken over a teaching position in the biology department at the university the year before, and his office was buried deep within the belly of the old complex. Savage found his way to the warm glow of Scott's bunker, and they began to look over their plans for the field season, big topographical maps spread out over Scott's desk in the dim light. But it wasn't long before their conversation was interrupted by a soft knock at the door, and they turned around to find Jerry James standing at the threshold with a glass jar underneath his arm.

James was known to drop by from time to time with descriptions of the amphibians he had seen among the cloud forests around Monteverde, and it was not unusual for him to appear at Scott's office with a new

---

* Cloud forest: A wet tropical mountain forest characterized by the presence of mist and cloud.

story to report. But Savage's attention was not focused on the young man from the Quaker community; his eyes were drawn to the shadowy bulk beneath James's arm and its unknown contents—something moving in the jar.

When James revealed his treasure to the two biologists, jar set down on the large desk in the light, Scott and Savage were struck silent by the iridescent toad contained within. For a long spell they said nothing, watching the *Bufo*[*] grasp against the slick glass of the capsule that had carried it so far away from home.

After taking in the sight, Savage asked James if somebody had dipped the toad in orange paint. He had never seen a toad so ornately decorated; nearly every other member of the genus was colored so it might disappear into the leaf litter when required—nothing like the toad in the jar before him now, which was gold and brilliant and looked as if it had been carried into the city from another world. Somehow, it had been overlooked in the mortal judgments of evolution. But James denied the charge of forgery, and Savage stooped down to observe the toad more closely, enraptured by the peculiar sight as James retold his story of discovering the species on the primitive ridgeline above the Quaker colony high in the mountains of Monteverde.

Savage looked from the extraordinary orange toad in the hazy jar to James's face, and, in the quiet of the dusky room, gold began to glimmer in his eyes.

With Jerry James as their guide, it took Jay Savage and Norman Scott close to twelve hours to make their way up the muddy road to Monteverde. They left the city at one o'clock in the morning, taking two jeeps—one to pull the other out when it became stuck in the deep mud of the steep road. Ten years earlier, a government tractor had worked to improve part of the route, but the main road branching off from the Pan American Highway at sea level was still steep and slick when the biologists

---

[*] *Bufo*: A genus of true toads in the family *Bufonidae*.

PROLOGUE

turned up into the mountains in the blue light of a false dawn. They followed the same path that the first Quakers had traveled more than a decade earlier, yellow eyes watching them from the shadows of the forest; in those days, the tigres[*] had not yet been hunted out of Monteverde.

At last, as the orange sun rose and they climbed into the highlands, they could see across the San Luis Valley to the rolling ridgeline of the Continental Divide. The thick green jungle had been cleared occasionally on the lower elevations for pasture; but on the steeper slopes, swallowed in some places by drifting mist, the old-growth forest was undisturbed and still largely unexplored.

It was early afternoon when they passed through the small towns of Santa Elena and Cerro Plano to reach Monteverde proper. There, a dirt road twisted between hidden homesteads swallowed by the forest, and dairy cows paused to watch the jeeps roll through, uncertain of their purpose. The cheese factory sat at a crossroads, one track leading off into green pasture, another down to ford the quebrada[†] and then up through dark tunnels into deeper forest. The Monteverde schoolhouse, looking small beneath the shadow of a tall fig tree, marked the boundary of the primary community. A few hundred meters past the home of Walter and Mary James, the vehicle road came to an end. There, Savage, Scott, and James abandoned the jeeps and trekked through wild pasture until they reached the great labyrinth of the cloud forest.

Savage's field notes and expedition report chronicle the trio's journey in detail, into a primeval forest prone to the elements and cloaked in secrets:

> *To reach the type locality of the new* Bufo, *we walked along a muddy meandering trail through the rainforest zone. After a hike of an hour*

---

[*] Colloquially, "tigre" would refer to a large cat. All six of Costa Rica's cat species have historically occurred in Monteverde: the ocelot, the margay, the oncilla, the puma, the jagarundi, and the jaguar. The last jaguar in Monteverde was shot on Marvin Rockwell's farm by Rafael Arguedas and Iginio Santamaría in November 1952.

[†] Quebrada: A stream.

PROLOGUE

and a half we came to another trail along the Continental Divide at an elevation of 1590–1600 m. Because of the constant prevailing winds from the Caribbean that blow across the divide from the northeast, the trees on the ridge are somewhat stunted and the forest is almost always shrouded in clouds or fog and receives heavy rains almost daily. Under the forest canopy of broad-leaved evergreen trees the limbs and trunks are dark in color and covered with a rich development of epiphytes.* The soil is black and covered with fallen leaves in various stages of decomposition. The general impression is of a deep, dark, dank, dripping woods, rather cool in temperature. During the day the enclosed forest has an eerie atmosphere created by the black soil, trunks, and tangled roots, and the billowing clouds of fog pass overhead through the trees. At night the interior of the forest is pitch black.

Among their field notes are hand-drawn maps of the machete trail that they followed up to the Continental Divide, collection notations, and Norman Scott's reference to their first sighting of the ephemeral and long-sought species: "A few days ago I entered in the books an orange *Bufo*." But even if they had come to believe everything that Jerry James had told them in the city, they would not have been prepared for the sight that they encountered when they reached the breeding grounds of the glimmering toads.

Climbing finally to the wind-battered ridgeline, wet to the bone and closed in by the dark and grasping hands of the elfin forest,† they stopped in their tracks and looked with wide eyes over an understory that was suddenly alive with color, motion, and endemic‡ magic. Collected around shallow pools of rainwater that had gathered in the

---

* Epiphytes: Plants that grow on other plants, absorbing water vapor from the air.

† Elfin forest: Also known as a *dwarf forest* or *bosque enano*, the elfin forest takes its name from the stunted and twisted trees bent and gnarled by the wind; its appearance is that of a forest of elves or dwarves.

‡ An endemic species is a species that is only found in one specific place, completely unique to that area.

root hollows and small depressions in the bedrock, a sweeping congregation of the orange toads lit the misty forest like a hundred little fires. Some peered out at the encroachers from hidden crannies in fallen trees, while others slept among broad green leaves in the musty leaf litter; mating pairs clasped together in the wavering pools were reflected in the dark mirror, their moving images bowing with the ripples in the celestial water.

"Everywhere, there were orange toads," Savage remembers. "Like a big explosion. There were hundreds of them."

In his field notes, Scott describes an "orange *Bufo* forest," intruded upon by a light mist, high on the wet ridge along the backbone of the earth. In an otherwise dark and cold and unforgiving ecosystem, the toads before them seemed impossible—yet here they were, alive and unlike anything that had been described before, crawling in and out of the low black pools in a forest that lay at the cusp of history.

For two hours, the biologists explored the stunted forest, documenting several hundred orange toads in an area no more than two hundred yards long and one hundred yards wide. They discovered that the females were a darker black, dotted with red and yellow blemishes, and were not as numerous as the golden males. In some pools, they encountered a dozen males wrestling for the attention of a single female, grasping over one another in a desperate bid to beat mortality. They collected many in great glass jars.

When the sun began to set—turning the orange toads into luminaries, burning in the dark—they retreated to a small shack in which Jerry James had often sheltered on his explorations through the elfin forest, farther east along the ridge above fifteen hundred meters. That night, they slept on the damp wood boards and listened to the wind howl up from the Atlantic slope, a cold rain washing over the enchanted mountain. In the dark corners of the shack, the toads in the collection jars ran down like old clocks in the night and, as the silver moon rose high above the shrouded forest, were finally still.

PROLOGUE

Over the next several days, Savage and Scott returned to the breeding grounds to observe and collect more toads; the expedition's field notes and subsequent reports indicate that 174 specimens were collected, along with an unrecorded number of eggs and tadpoles.

When a Kansas herpetologist named Ed Taylor visited a few weeks later, Savage revealed one of the golden males that he had kept alive in a transparent jug. Taylor, who had traveled extensively throughout the country, remarked: "The good lord has been busy since I was last in Costa Rica."

In his paper announcing the documentation of the new species—"An Extraordinary New Toad (*Bufo*) from Costa Rica," published in 1965—Jay Savage acknowledged Jerry James as the first person to encounter the toads, and thanked Norman Scott for his field assistance, but history has remembered Savage as the primary discoverer of the novel species. In the same paper, Savage christened the toad as *Bufo periglenes*, "from the Greek *periglenes*, meaning very bright." The species would come to be commonly known around the world as "the golden toad," though in Monteverde it was often referred to by another name—one that had a long history in the forests of Central and South America. Like the myth of the golden city that had conjured dreams of fortune and glory—drawing explorers and conquerors into ruin—the golden toad was locally known as "el sapo dorado," or simply "el dorado."

The first pair that Savage brought back to the United States, male and female, are still preserved at the Natural History Museum of Los Angeles County. The other specimens went out like emissaries to the major world collections in Paris, Great Britain, Berlin. There they remain, the last remnants of a vanished species, with no voices left to tell their story—no answers to the mystery that they left behind when they disappeared for the last time beneath the earth, never to return.

CHAPTER ONE

GHOST STORIES

Monteverde, Costa Rica

In the spring of 1987, six years before my twin brother, Kyle, and I were born, our father brought a ghost home from the tropics. He had traveled to Costa Rica with our mother in the moody tempests of the rainy season, and they had climbed the green hills into Monteverde on a bus that teetered on the edges of eroding cliffs. Like other ghost stories you might have heard, theirs began with a summons from an old friend to travel to a hidden forest, far away from what they knew.

The invitation had come from Alan Masters, a former classmate of our father's at the University of Florida. When they had been in school together, Alan had studied monarch butterflies, and our father had studied viceroys; their paths had often crossed as they chased the orange and black butterflies across the prairies of central Florida or along the median of Highway 441, alligators eyeing them from the banks of long canals. They might have become competitors, but instead they became friends. When Alan and his wife, Karen, moved to the small town of Monteverde to spend six months researching neotropical butterflies, Alan wrote to his old friends:

> *Karen and I are currently rearing tons of butterflies. Karen has started a new project on fig wasps so our house is full of them. They're really interesting to watch but in our mashed potatoes?! I guess it's good to get a little meat with your potatoes—it's not steak but we are (supposedly)*

roughing it here in THE TROPICS. *Well, have you guys made your airline reservations for Costa Rica yet? Keep us posted of your travel plans.*

When they joined the Masters in the high hills of Monteverde, our parents found a small community that had grown up around the seeds of the Quaker settlements and the local homesteaders: a Catholic church on the square in Santa Elena, a few small hostels, the lechería*, and the Friends school farther up the hill in Monteverde proper. Though still modest, there was no pretending that it was the same town Jay Savage had visited twenty-three years earlier; by then, the secret was out. The golden toads had put Monteverde on the map, and droves of biologists and collectors had been ascending the mountain over the last decade in a pilgrimage to explore, document, and sometimes plunder the region's unprecedented biodiversity. Karen and Alan Masters had arrived in Monteverde at the dawning of a new era, when the locals who had once made their livings cutting down trees were beginning to develop a more symbiotic relationship with the cloud forest. With scribbled directions from an international call, my parents made their way through the changing town and met the Masters at their little house folded in the deep hills.

When the morning sun rose over the forest, Karen set off for her field work in the dry valley of San Luis, catching a ride on the milk truck descending the precipitous trocha† from the cheese factory, while Alan led our parents up the steep trails through the twisted forests on the mountain's towering ridge. Alan Masters had not traveled to Monteverde to study the golden toads, but he knew them well enough, and he told the story of their discovery to our parents as they explored the winding paths of the Reserve. Monteverde was home to a staggering diversity of species, but the golden toads were special: This

---

* Lechería: The Quaker dairy factory.

† Trocha: A steep, long, narrow path. "The trocha" is the local name for the steep road from San Luis to Monteverde.

was the only mountain in the world where they were known to live; they were bright and enigmatic, while their closest relatives were plain and modest; and they only appeared for a few days every year, disappearing into unknown sanctums when the rains ended. They were a perennial reminder that for everything the biologists had discovered in Monteverde, there were still more mysteries. Alan led our parents through the untamed and rugged woods on the highline of the mountain, but they never saw the golden toads. That season, the last of the toads had descended beneath the earth at the end of May: Our parents had only missed them by a few weeks.

But they encountered other treasures in the remote forests of Monteverde. They found huge brown toads in the long grasses around the lechería, followed dancing blue morpho butterflies into the hills, and watched the burning sunsets over the distant water from the boulders at the forest's edge. Each night, our father would creep out to the porch to see the moths that had braved the wind and rain to come and worship the orange glow of the deck light in the dark. To him, the tropics must have seemed too good to be true: an oasis amid a crueler world that was known to kill its tenants. In any case, it would be five years before he would return, and by then the golden toads were long gone.

When he watched the transient tropical moths flying in to meet the buzzing porch lights on the Masters' deck, our father might have seen himself within them—like a distorted reflection in the glass of a bulb. He'd been diagnosed with a cardiac condition when he was only eleven years old; the specialist at the big hospital in Madison told him that his heart could stop at any moment. It was a prescription to live in fear, beneath the shadow of death. The spring after his diagnosis, he began to raise moths in his bedroom, hiding the precious cocoons from his mother like contraband treasures. He learned how short the life of a moth could be: The luna moth only lived for seven days, and it didn't even have a mouth—but it spent its short life dancing in the silver glow of the moon anyway.

This was a ritual that he continued across his adolescence, through adulthood, and into parenthood. When my brother and I were young, the house was filled with the sweet smell of willow leaves and caterpillars that our father carried in to spare them from the cold and frost of late November. The walls of his office were lined with bright specimens in display cases, the oddments and mementos of his research suspended forever behind clear glass. The bright colors of long-gone butterflies were preserved in the stacks of photo slides that he kept beside an old projector in a box, dust rising in the flickering light like spirits ascending from the dark.

On rare occasions, he would lift that mechanical relic from its hibernation; to prepare a lecture for the classes that he taught at the local college, or when an uncharacteristic spell of nostalgia descended from the unknown regions. On those nights, he would turn the old contraption on—its clicks and purrs sounding like a dreaming animal—and we would crawl into the boundary of its light while he told stories illustrated by the dancing images projected on a sheet. This happened once when Kyle and I were eight years old, in the den of our house at night. He unpacked the slides from their black coffins and slipped them into the carousel, and the dark room was painted with immortal images, memories preserved on film. There were photographs from his childhood in cold Wisconsin, which he clicked through at the pace that one might leave a haunted house. There were pictures of gleaming salamanders hidden in the dark caves of Appalachian waterfalls. There were portraits of the hot southwestern deserts he'd explored as a teenager—diamondback rattlesnakes, and saguaros like penitents frozen with their arms in the sky. And there were the images that he had carried back from the tropics: delicate clear-wings and painted *heliconius* butterflies, little jewels among the enchanted forest.

And then the dark room turned bright orange, lit up by the opulent glow of the golden toad. The photograph had captured a bright male emerging from a rain pool, a spark of color in the brown woods and black water; mist swirled in the dripping forest, where gathering kin glowed like little pools of melted gold.

It was the first time that I had ever seen the golden toads, and I knew nothing more than what our father told us: that they only ever lived at the top of one mountain in the cloud forest, that they used to gather in the hundreds when the rains swept over Monteverde, that they were gone.

I don't know which was more fascinating to me in that moment: that such a strange and beautiful toad could exist in distant forests, or that it had vanished without a trace. The fact that he had never seen them for himself did not diminish the quality of his eulogy. The photo slide must have come from Karen Masters, or from one of the other Monteverde biologists who had climbed the green mountain during the first rains, but he kept it among the other treasures of his own, the talisman of a lost world that had once existed and was now gone. It was a ghost he carried home from the mountain wrapped in mist, but a ghost that he pitied, and he spoke about it with a certain sympathy: The doctors had told him that he could vanish just as easily. So we sat in the shimmer of the orange glow, haunted in the dark, and felt the spell of the golden toads settle over us.

When he raised the window shade, the mystery disappeared in the light. Piece by piece, Kyle and I watched our father dismantle the old projector and lay it back into its sarcophagus. He slipped the photo slide back among the others, where it was lost somewhere among the mementos of the past. But the golden toad burrowed deep into the soil of my memory, where it dug itself a warren, lay still, and was forgotten for many years.

San José, Costa Rica

That spirit didn't begin to stir again until I wandered into the forests that it haunted. In August 2015, when I was twenty-two years old, I saw

Costa Rica for the first time from above. The landmarks that would grow familiar were still mysterious and unknown, and I could only guess in which direction lay the green mountain that would be my ultimate destination. On the busy streets outside the Juan Santamaría International Airport, a towering Costa Rican biologist called Moncho leaned out of an idling van to wave me over, and I piled into the back seat with two other interns as we pulled out onto the Pan American Highway, heading for Monteverde.

Like my parents, I had gone into the tropics at the invitation of Karen and Alan Masters. Kyle and I had spent the last four years living together in a cramped dorm room at a small college in South Carolina, and by the summer of 2015, South Carolina had worn out its welcome with us, and we had worn out our welcomes with each other. Kyle was going west to California for graduate school, but I had been relegated to the wait lists of all the graduate programs I'd applied to. One night, searching the ECOLOG listserv for interesting jobs that would get me out of the Carolinas, I found an opening for a teaching assistant with the Masters' field program based in Monteverde. I suspected that I wasn't qualified, and they agreed, but they responded to my email with a counteroffer: If I would fly to Costa Rica in August and spend two semesters taking photos and writing a blog for their study-abroad programs, they would lodge me and feed me during my nine months in the cloud forest. I took the deal without a second thought, Ben Loman's words from *Death of a Salesman* beckoning me on: "The jungle is dark but full of diamonds. One must go in to fetch a diamond out." While Kyle packed the car for a long drive into the West, I packed my bags to go into the jungle. It would be the first time we would be apart for more than a few weeks since we'd been born. My expectations of the tropics were informed only by the stories I had heard, colored by our father's occasional and perplexing advice: "Remember that everything you touch there can kill you. You'll be fine."

GHOST STORIES

It must have been hard for him to watch us go—me in one direction, Kyle in the other. In his young adulthood, the family that our father had grown up in had collapsed around him. He had lost his younger brother and then his sister, two out of his three siblings, and his father died when Kyle and I were eight years old; he hadn't traveled to Wisconsin to see his mother in years. From the fragments of his past, he had built himself a new family: our mother, Kyle, and me. He gave us everything that he had missed or lost when the family that he'd known began to crumble: attention, understanding, unconditional support. Now we were rewarding him by leaving. He would have to have faith in us; that the paths we were following would lead us back someday, or that there would be a place for him in the unknowable futures we were building far from home.

Unconcerned with these complexities, I spent the weeks leading up to my departure sorting through the piles of gear I had amassed and reading what I could about the ecology and culture of Costa Rica. The day before I was scheduled to depart, Karen Masters called me and offered one last piece of mystifying counsel: "You should know that Central America is the land of magical realism. One minute a person will be walking down the street, and the next they'll be floating." After that, I gave up the last of my half-hearted preparations.

I heard my first Costa Rican ghost story on my first evening in Monteverde. On the drive up from San José, between fitful spells of disoriented sleep, I watched the city fade into green fields, canyons, thick hillsides, dry forest, and finally the coast. After a few hours, we turned away from the sandy shoulders where carloads of children played in the surf and suntanned vendors sold street meat and granizado[*] from metal carts, and the van went up the steep road that climbed into the mountains. The sun sank into the water on our slow ascent, and the buzz of night insects arose in the grassy ditches by the road—the music of a

---

[*] Granizado: A Costa Rican dessert made of shaved ice and topped with condensed and powdered milk.

tropical evening. It was after dark when we reached the biological station at the top of the mountain, and everyone was tired. From the top of the hill, we could see the yellow lights of Santa Elena glimmering below, but at this altitude—in this solitude—it was a different world. Moncho said good night and descended to his cabin hidden on the slope, leaving me with the second field assistant who had ridden up with us from the city: Raquel, a small and unassuming Tica* who wore a hand-painted jaguar's face on her shirt beneath a denim jacket. She flipped through a ring of keys as she led me up the winding hill to my little bungalow in the shadow of the forest. To fill the silence, she told me ghost stories.

"Not all of these are real," she clarified, "but some of them might be."

Far below us down the steep walls of the valley, the trickle of the little waterfall sounded like murmuring voices sharing secrets, or offering curses in the dark.

"There is one," she said. "Do you know la segua? She is, like, a beautiful woman you hear crying in the forest. Like—if you are drunk, or you are walking by yourself at night. Her body is a beautiful woman, but she has a horse's head—with red eyes."

The wind kicked up, sending the canopy of trees swaying to caress each other, passing warnings. The running water in the black valley began to sing with the voice of an enchantress.

"But if you go in to try to help her, you disappear."

She was walking with her head down, still sifting through the ring of keys. The night's errand, for her, was nearly over. Around us, the cloud forest closed in. It was a forest where things disappeared.

"And that one, I think, is definitely real," she said. She found the key that would open my lonely bungalow and took it off the ring, holding it out to me like a charm. In the moonlight, it was white as bone.

"Bueno, good night," she offered with a kind smile when I took the key, then turned and disappeared down the steep hill into the blackberry

---

* Tica/Tico: A local term for a person of Costa Rican nationality.

night, leaving me to finish the eerie walk through the restless jungle alone.

That night, I wrote my first letters home from the bungalow in the shadow of the misty forest. The first, I addressed to Kyle's new apartment in Riverside, California; I knew that he would be camping in the Mojave Desert on his long drive to the West Coast, and I wondered if my letter would reach his new home before he did. He had plans to light off for Santa Cruz Island soon, in search of the Channel Island Foxes that had recently clawed their way back from the edges of extinction. The second, I addressed to my parents' house in South Carolina, telling them everything I'd seen so far, and all the ways the town of Monteverde had changed since they had seen it last, almost thirty years ago.

In the morning, sunrise chased off Raquel's ghost stories and I walked the letters down to the post office in Santa Elena, where I counted out a few hundred colones* to send them north, through jungles and deserts, into the lands that I'd departed.

I spent my first week preparing for the arrival of a dozen students from the United States who would spend the semester in Monteverde as part of the Masters' field program. In the early-morning mist, I followed strange maps to find the hidden paths that climbed to Cerro Amigos in the cloud forest above the biological station, where Karen had hidden camera traps near the puma trails and where a local biologist had just found tapir tracks in the soft mud along the ridge—the first time they'd been seen in years. I scared up agoutis and coatis and black guans (the older campesinos called these meaty birds "flying soup") as I slipped down hillsides following cold streams to waterfalls, getting lost in the high and stunted forests at the mercy of a godly wind.

When the rains descended during sleepy afternoons, I explored the biological station. Past the offices in long halls smelling of formaldehyde, I found cabinets of preserved butterflies—electric blue or with delicate

---

* Colones: The currency of Costa Rica.

transparent wings—and rooms full of microscopes and centrifuges. There was a large classroom at one end of the station's second floor, beside a library filled with identification guides and histories of the tropics and perplexing field-medicine manuals. (Werner's *Where There Is No Doctor* recommended: "If you have a strange sickness, do not blame a witch; it is impossible to bewitch a person who does not believe in witchcraft. Do not seek revenge against a witch because it will not solve anything.")

On one of those afternoons, I heard my second Costa Rican ghost story. In a dark closet beside the library, I discovered a small poster hidden among the cobwebs. It was a desperado-style handbill bearing the title "Wanted Alive." Below the text were the renderings of ten lost amphibians. Some of them were color photographs, and others were black and white or hand-drawn; many of the species had not been seen in a century, and some had never been photographed. I would learn later that the poster was a part of Conservation International's 2010 "Search for Lost Frogs" campaign, but at that time I had never heard of Conservation International, and I didn't know that lost frogs might be rediscovered. That afternoon, all of my attention was focused on the solitary species at the top of the poster, the #1 Most Wanted: the golden toad. Beside the photo of the bright orange toad on a wet green leaf was a line of bold text: "One rainy evening in May 1989, a lone golden toad appeared at a pool high in a Costa Rican cloud forest. He was the last golden toad ever seen."

In a memory grown foggy through the years, dirt shifted on a forgotten grave. I saw again the shorn fragments in flecks of color: the yellow of the old projector's blinking glow, the green of the elfin forest tangled on the ridge, the gold of the mythical toads converging in black pools. It had been close to fifteen years since I had last thought about the golden toad, and now I found myself by accident in the forests that it had once called home.

Carefully, I pulled the thumbtacks out of the wood and carried the poster down the hall to Moncho. I found him sitting in the office on the second floor of the station, where the big window looked out on the emerald valley falling away into the pale blue line of the gulf. When he saw me, he raised his mug and pointed at the fridge beside his desk—an invitation to make myself a drink—and I asked him what he knew about the golden toad.

It was a question he had likely heard from a dozen students over his last ten years working with the Masters, and he told the story like an old folk tale: Savage's discovery in 1964, the boom, the ultimate decline. I stayed and had a drink with him—our last before the students would arrive in San José in the hazy predawn of the following morning—and thought about how strange it was for a whole existence to be abridged into a story you could tell in one rainy afternoon.

The fog blew in, blotting out the view of the gulf. The other field assistants wandered in and out of the library down the hall, collecting textbooks that we would pile into boxes for the students flying in to San José. Moncho and I sat in the office for a while, drinking rum in metallic mugs and watching the coatis dash in and out of the sheltering forest in the rain. Soon my thoughts had wandered from the golden toad, and at first I wasn't certain what Moncho meant when he added in a vague notation, "But you know, that poster is wrong."

He was leaning forward to open the fridge beside his desk, one finger pointing in my direction. I remembered that I was still holding the "Wanted Alive" poster, so I raised it in the light and examined it more closely, though it looked all right to me.

I asked him what he meant, and he replied, "Well, it is depending on who you ask," as he poured another few fingers of flor de caña into his cup. "But there is one person who will say that he saw it later, at another place."

In the hollow of the windowsill, a spider caught a moth in its ragged web while heavy beads of rain ran down the glass. I tried to remember the way my father had told the story. I was sure that he had said that the last golden toad had disappeared into the forests above Monteverde in

1989, a last survivor on a lonely ridge, its kind never to be seen again. In my college textbooks, the story had been the same: one toad seen in 1989 by an American biologist, then never again. This was the first exception to the standard history I had ever encountered, the first impression that the story of the golden toad's extinction was not so simple.

Outside, the August rain was turning the forest a deep, dark green, and it was not so difficult to imagine an orange toad emerging from hidden places like a refugee from a great calamity. I asked Moncho who this person was, who had seen the ghost so long after its retreat, and Moncho answered: "Eladio Cruz."

Peñas Blancas, Costa Rica

It was two months in Monteverde before I crossed paths with Eladio. That October, I was struggling along a steep mud bank on the Atlantic slope of the Continental Divide, and Eladio Cruz was somewhere up ahead of me on the slick trail slinking down into the Peñas Blancas Valley. He was leading Alan Masters's group of students down into the wild forests to spend a few days at the old homestead he had carved out of the jungle, where the students would look for tree frogs and eyelash vipers and the creeping mushrooms that grew along the stone banks of the spitting river. It was the rainy season, and the way down into the valley was wet and overgrown. It took us eight or nine hours to climb to the ridge above Monteverde and drop down the other side, crossing old suspension bridges over canyons and following the footprints of Eladio's white packhorse through the deep mud, ten miles into Peñas.

By the time I reached the homestead in the valley, the evening was a gray-blue haze and the sun had disappeared into the sea of trees, the rain blowing in only a few minutes after I'd arrived. The horseflies were just beginning to emerge when I found Eladio in the tall grass behind the house, unloading bags of rice from the horse's saddle. He was sixty-seven, and he had reached the bottom of the valley in half the time it had taken me. In the fading light, he was only a small, dark shape in a deep jungle.

That night, Moncho translated Eladio's stories to the students over candlelight in the kitchen of the old house—coming down into the valley for the first time with his father carrying banana seeds, or the time that he was bitten by a fer-de-lance*—while the forest rangers drank clear cacique on the porch and the horseflies buzzed into the flickering halos, until they were caught in the candle wax and trapped forever. At night, the forest came alive with strange songs and chatters and once, a little after midnight, a low growl that brought a waiting silence until the thin hours of the early morning.

In those five days we spent in the valley, Eladio never talked about the golden toads. He talked about the orchids that grew on decaying branches in the woods up and down the slopes of the cordillera, and about the strong rivers that cut deep lines in the valley, sometimes overflowing and sending dark water into the lowland rainforests. He reached into thick leaves and drew out glittering frogs, and he cleared old trails climbing up into the highlands with his long machete. But even in the slow evenings when the rice was cooking over the big fire in the kitchen, he told no tales about the golden toads, and I wasn't brave enough to ask him. I couldn't speak his language very well, and I was afraid that he might answer my question in ways that I could not understand. Or maybe he would decline to answer me at all. He didn't know me very well, and these were local secrets. Why should he reveal his ghost stories to a stranger? So I was left wondering, and the mystery remained.

It wasn't until our last night in the valley that Eladio told a story about a frog—though it was not the one that I had been anticipating. In his usual shy and humble style (and after some encouragement from Moncho), Eladio described his rediscovery of the red-eyed leaf frog, *Agalychnis callidryas*, a species that the scientific community had

---

* Fer-de-lance: Known locally as the terciopelo, the fer-de-lance snake (*Bothrops asper*) is "the most feared and dangerous in Central America."

believed to be locally extinct in the Peñas Blancas Valley for years. At first, the biologists in Monteverde had not believed him; but, hiking down into the valley's sheltered forest, they had borne witness to the revenants for themselves. As Eladio told the story in the shadow of the very jungle that had shielded the survivors, I understood for the first time how deep the forests surrounding Monteverde truly were—and how easily a lost species might slip back into the dark, to go on breathing but never to be seen again.

The next morning, we hiked out of the Peñas Blancas Valley and through the vast protected forest on the Atlantic slope into Poco Sol, where we would stay for a few days before heading south to Panama's Bocas del Toro. It was on a beach in Bocas that the old biologist Richard LaVal met the group and told the story of when he had rediscovered "the first of the so-called extinct bats." That was the second of several accounts that I would go on to hear of resurrection in the tropics.

I had signed on for nine months in Monteverde, but in the end I stayed in Costa Rica for two years. I made my way to the Osa Peninsula, where I met a silent puma, ghost-cat, crossing the path ahead of me between the forest and the beach. I climbed to Cerro de la Muerte—the mountain of death—to search for salamanders, mountain tapirs, and little spirits they called nímbulos, the children of the clouds. I crossed the old frontera by the migrant roads into Nicaragua to photograph the scars of deforestation, ancient forests brought down for cattle pastures, and there were ghosts there too—La Sombra, who stalked the shadows around the collapsing fortress at El Castillo, the nighttime echoes of cannon fire from a long-ago battle, smoke rising from the sugarcane fields on fire, animals vanishing in the disappearing forest. In Monteverde, I fell in love with the woman who would become my wife, and I fell in love with Monteverde too. I began to feel at home. From time to time, I crossed paths with Eladio or saw him from afar as he followed the old trails. He began to recognize me, but I wasn't sure he knew my name.

GHOST STORIES

In August 2017, I boarded a plane and moved back to the United States. *When I was twenty-two, I walked into the jungle*, I thought to myself as the green lands grew small beneath me, *and when I was twenty-four I walked out—and, by God, I was rich.* In those years in the jungle, I had found my wife, a wild and imperiled cloud forest, and endemic ghost stories with deep roots in the places they had grown from—better riches than I had any right to. But among the legends piled like treasure, one jewel glinted brighter than the others. As the country disappeared into a shroud of fog, I was sure that I'd return. My dreams on that long flight were haunted by the glimmer of the golden toads, and I awoke with the sensation of sitting at the brink of an eternal forest, where death might be suspended and the stories might be true.

CHAPTER TWO
THE GOLDEN AGE AND THE GREEN MOUNTAIN

Fairhope, Alabama

Before science brought the first biologists to Monteverde, religion brought the Quakers. In the winter of 1949, four Quaker dairy farmers sat in the Baldwin County Courthouse before U.S. District Judge John McDuffie, waiting to be sentenced on a charge of failing to comply with the military draft registration. Wilford "Wolf" Guindon, Marvin Rockwell, Howard Rockwell, and Leonard Rockwell had thought the matter over and determined that they could no longer in good conscience support the war economy of the United States, nor register for the draft at risk of indirectly sanctioning war. Like the early members of the Quaker religion who had fled England and religious persecution to land on North American shores, they still held to one of the key tenets of their faith: the belief that war and violence were against the will of God. Though some of the Fairhope Quakers had compromised by necessity on this statute in the past, a small number of their community felt compelled to take a moral stand, and they sent letters to the draft board in Foley, Alabama, to explain their motivations for protesting the registration. They were arrested and brought to trial; but over the course of the depositions, Judge McDuffie found himself growing sympathetic to their circumstances.

"The world has been hoping for world peace," McDuffie remarked, "and it never will happen until God changes the human heart."

A supporter of the defendants replied: "But if we have a few human hearts that have been changed, should they be punished?"

THE GOLDEN TOAD

McDuffie was caught in a troublesome dilemma: His wife was descended from the Pennsylvania Quakers, so he felt an ache of compassion at the plight of the young men who had been brought before him for judgment. But he was also bound by his professional ethics and a duty to uphold the rule of law. He could not easily make exceptions when his personal feelings were at odds with the edicts of the nation.

"The law is: You must register," he eventually determined. "There is nothing I can do but to sentence you unless I go back on my oath of office." Before the sentencing, however, McDuffie handed down another statement—one that would prove to be even more significant in the trajectory of the defendants' futures: "If you like this country, you should obey the laws of this country, and if you don't like it, you ought to move out."

After a formal plea of *nolo contendere**, the four Quakers were sentenced to one year and one day in prison; they were remanded to the federal penitentiary in Tallahassee, Florida, where they served four months of their sentence before being released on parole. Through their long days behind bars, the Quakers had time to turn over the judge's parting words. By the time they stepped once more into sunlit freedom, salty air blowing up from the Gulf of Mexico, they and others in their community had determined that the time had come to leave the United States and—like their ancestors who had fled from Europe—seek a new home where they might build a better world.

They ruled out frigid Canada fairly quickly, as the freezing temperatures of the neighboring country would prohibit them from making their livings as farmers. They considered Australia, but it was too far from their families and neighbors; visits back to the United States would be difficult and costly, if not impossible. Mexico was disqualified due

---

* Nolo contendere: A plea of "no contest" in which the defendant accepts conviction without officially admitting guilt. It is, memorably, the same plea that the authors' father submitted when criminally charged with stealing pumpkins outside of Madison, Wisconsin, in 1977.

to the country's restrictions on land ownership. The Central American isthmus, nestled between two worlds, appeared to be their final hope.

Embarking into the unknown regions, the Quakers were particularly impressed with Costa Rica. With a stable government, sound economy, friendly people, and farmable land, the small country seemed to be what they were searching for. And there was another enticing detail: After a violent civil war, the government had abolished the Costa Rican military in 1948, making it one of the only countries in the world without a standing army. For the Quakers fleeing persecution for their pacifism, it seemed to be a sign from God.

When the first of the expatriates touched down at the Sabana airfield (the airport that Jay Savage would fly into a decade later had yet to be constructed), they began a survey of land for sale near Cañas Gordas close to the Panama border. Others explored northwest toward Los Chiles and Caño Negro, or southeast to San Isidro, riding horses up into the mountains from the edges of small towns, but the land they found was too expensive, too isolated, too cold; in the lowlands they met disease and debilitating heat, impassable forest, soil ill-suited for farming. Christmas came and went, and the search for land continued, becoming desperate.

In the spring of 1951, with the rainy season looming, news came to the group of one last prospect: an area high in the Tilarán Mountains northwest of San José, which contained a shelf of level terrain among slopes of primary cloud forest. The Guacimal Land Company currently held the title to the region, but they were open to selling a tract to the Quakers if the land suited their needs. Cautiously optimistic, three envoys traveled to Guacimal in the province of Puntarenas, where the Pan American Highway degraded to a primitive dirt road pointing up into the mountains, and they set off on horseback up a steep and rocky pass that had never seen anything but oxcarts and horses. On the lower slopes, much of the land had been cleared or burned for pastures—some patches still smoldered—but as they climbed higher above the canopies, traversing wooden bridges and rolling ridge-tops, they found themselves peering through the fog into a different world.

The cool air and cloud vapor drifting over the deep green mountains washed the forest in a nurturing mist, infusing the soil with the vital moisture that the wandering farmers had been searching for. Uncut forest stretched out of sight on the steep slopes, but they spied flatter areas where they could raise cattle among the sheltering windbreaks and erect small houses to begin their new community. From the plateaus, they could look down through the mist and the smoke from lingering fires to the pale blue waters of the Nicoya Gulf.

On this remote mountain, they would have plenty of room to grow. It was clear that Indigenous clans had seasonally occupied the lower elevations, leaving behind pottery and corn-grinding stones, but only within the last generation had Costa Rican homesteaders begun colonizing the thick cloud forest closer to the Continental Divide. Gold had drawn the first contemporary settlers to the mountain; from the Pacific slope, gold miners from Guacimal hiked up into the cloud forests to hunt, sometimes building small shelters for extended trips. By the time the North Americans arrived, there were close to a hundred subsistence farms in the area belonging to families with names like Arguedas, Leitón, Méndez, and Villalobos, selling food and guaro[*] to the miners in Guacimal and Las Juntas. The far side of the mountain was less domesticated, where rain and mist washed the forest in a perpetual sea of fog. The Quakers heard stories of "a small trail leading up to the Continental Divide and then down into a mysterious place called 'Peñas Blancas' where 'wild people' lived and where there might even be gold to be found."

As part of their agreement with the Guacimal Land Company, the Quakers would pay the Costa Rican homesteaders for their properties and purchase the titles from the Costa Rican government. The

---

[*] Guaro: A liquor made from sugarcane.

THE GOLDEN AGE AND THE GREEN MOUNTAIN

small town of Cerro Plano—hosting a school, a Catholic church, and a pulpería*—would remain with the local farmers, from whom the new pilgrims would learn much about the secrets of volcanic soils. They would come to call their new community on the green mountain *Monteverde*.

Finally satisfied, the three Quakers turned their horses southward and began the long journey back down the mountain in the dark, carrying the good news home to those who were waiting in the Central Valley; they had found a homeland hidden in the mist of the wandering cordillera, a place where they could send their desperate roots deep into the earth. By that time, Judge John McDuffie had died of a heart attack; he would never know that the Quakers had taken his advice, setting off into the wilds of another world far away from the Alabama hills that they had once called home.

Exactly when the Quakers first discovered the bright orange toads hiding among the misty dwarf forests of the green mountain remains unknown. Certainly by the early 1960s, Jerry James was familiar enough with the species to anticipate their emergence and capture a lonely specimen in a mason jar to show the visiting biologists; but that was more than ten years after the Quakers had first begun to settle Monteverde. If others in the community had encountered the toads in great numbers prior to James's explorations in the higher cloud forests, that history was never recorded. The earliest accounts of the bright and peculiar toads come from one of the four Quakers sentenced in the Baldwin County Courthouse: Wolf Guindon. After serving his sentence, Wolf joined the expatriate Quakers on their pilgrimage into the green hills of Monteverde. While his new wife, Lucky, watched their son and worked long days to turn their spartan quarters† into a home,

---

* Pulpería: A small general store.

† For the early years, their home was a fourteen-by-sixteen-foot canvas tent—one of several that the group had carried with them as temporary lodgings while they worked together to stand up homesteads.

THE GOLDEN TOAD

Wolf hunted game to bring back to the family's table and prowled the windbeaten Continental Divide in search of farmable land. He quickly garnered a reputation up and down the mountain as one who might emerge like a phantom from the thick curtain of forest, bellow a cheerful greeting, and then disappear back into the dark to find his path again. Ultimately, that path would lead him to the doorstep of Monteverde's most enduring mystery, hidden among the tangles of the lonesome elfin forest.

"He was developing a farm right in the center of golden toad habitat," Wolf's son Ricky remembers, "and we heard all about it. Ever since I can remember, since I was just a boy, I remember being shown the golden toads."

In partnership with his uncle Walter James, Wolf possessed a claim to a piece of forested land high on Monteverde's Continental Divide, where few of the other homesteaders had ventured since arriving on the mountain. It is not hard to imagine why they hadn't: At that altitude, the whole forest was caught in a mist of fog and rain, and the trees were bent and mangled by the winds. Far above the town and scattered farms, it was a cold and lonesome place. But Wolf cut himself a path into the stunted forest with his machete, determined to draw the boundaries of a pasture he intended to develop with his uncle. There, in the enchanted highlands that had turned back so many others before him, Wolf encountered the mysterious toads that lived among the roots and rain pools.

"They were a real dull golden color," he remembered later, "but in breeding-season they were brighter, almost red-orange." From what Wolf could tell, the toads seemed to appear at the onset of the first heavy rains in an area nearly five miles long on the backbone of the Continental Divide, but no more than a quarter of a mile wide. He observed that the males would emerge from their hidden burrows a few days before the females, in anticipation of their courtship when the heavenly water filled the pools.

THE GOLDEN AGE AND THE GREEN MOUNTAIN

"That's just about the only time we'd see them," he noted, "as they were hiding the rest of the year." And when Jerry James set out to collect a specimen to carry down to Jay Savage and Norman Scott in the city, it was Wolf who helped him capture the toad—a lone emissary of their secret kingdom.

But even before word of the ethereal orange toads had reached Jay Savage, the first intrepid biologists had ventured into Monteverde, and the Quaker colony at the top of the isolated mountain offered access to the highland cloud forests that would have otherwise been inaccessible. The first formal research project in Monteverde was undertaken just over a year before Jay Savage would observe the golden toads for the first time, when a biologist from the University of Kansas spent three weeks in Monteverde as part of his study on army ants and the camp-follower insects trailing the colonies. But even he was not the first biologist to reach the mountaintop: As early as 1962, an ornithologist had visited the Quaker colony and hiked up into the cloud forests above the town, reportedly sighting seventy different species of birds on his first afternoon alone.

As more and more biologists made the pilgrimage to the secluded mountain, they began to discover that the little jewel of Monteverde was one of the most biologically diverse ecosystems on Earth. Over the next two decades, biologists compiling local species lists would quantify Monteverde's biodiversity, tallying 60 amphibian species, 101 reptile species, 120 mammal species, 425 bird species, 658 butterfly species, and more than 3,200 species of plants (it is believed to be the site of the highest diversity of orchids on Earth, with more than 500 known species—34 of which were discovered, new to science, in Monteverde)[*]. Close to 50 percent of Costa Rica's biodiversity is found in the Monteverde area

---

[*] For comparison: Yosemite National Park boasts 11 amphibian species, 22 reptile species, 90 mammal species, 262 bird species, 100 butterfly species, and 1,450 plant species in a national park more than 1,000 square miles in size. The most generous estimates of the Monteverde area, on the other hand, put it at less than 100 square miles.

alone, and the region's cloud forest is estimated to contain 2.5 percent of the biodiversity of the entire planet. This staggering species richness is due to the fact that seven of Costa Rica's twelve life zones are represented in the forests around Monteverde. If you started from the bottom of the valley on the Pacific slope near the San Luis bridge and hiked into the upper canyon of the Río Guacimal through Bajo del Tigre, across the coffee farms on the outskirts of Santa Elena, into the tall dense cloud forest and past the waterfall coming down from Cerro Amigos, into the elfin forests up and over the Continental Divide, and followed the river down into Peñas Blancas to the rainforests around the foothills of Volcán Arenal, you would see the forest's microclimates change as you moved through it, from dry forest to mist-enshrouded highlands. And if you had crossed through that elfin forest as the first spring storms were thundering above the ridgeline, stopping long enough to walk the old trails that wandered between the breeding pools, you would have seen the endemic treasure of Monteverde: the golden toads.

Jay Savage's scientific description of the otherworldly golden toads put Monteverde on the map, signaling to both established and emerging researchers that there were still some secrets left to be unearthed in the cloud forests. Two years earlier, Savage had been instrumental in the cofounding of the Organization for Tropical Studies (OTS), a collaborative program between seven North American universities and the University of Costa Rica to develop ecology field programs for U.S. students. By the early 1970s, OTS was bringing groups of graduate students to Monteverde to undertake research projects; the first group slept in the loft of the local pensión,* and later groups of undergraduates camped in tents for weeks of study.

The blank edges of the maps of Monteverde began to be filled in, and the accumulating published research—like George Powell's report on local bird species for *American Birds* in the early 1970s—helped to

---

* Pensión: A small lodge, like a hostel.

put the word out to up-and-coming biologists, birders, and naturalists looking for a new world to explore. Ornithologists drove their pickup trucks from Cornell to Monteverde to study cloud forest hummingbirds, passing logging trucks on the dirt roads up the mountain; entomologists arrived to study mimicry in clearwing butterflies and became beguiled by tropical plants, remaining in Monteverde long after their initial projects were complete; mammalogists adopted nocturnal schedules to observe the midnight habits of Monteverde's bats; canopy researchers climbed into the treetops for a different perspective on the cloud forests, revealing a new world seen from above.

In the same way that Paris in the 1920s drew young writers and artists into its orbit, Monteverde in the 1970s was a meeting ground for expatriate biologists seeking new discoveries in the mysterious cloud forests. Beyond the diversity of species and the variety of life zones, two other factors helped to make Monteverde one of the most sought-after study sites for visiting scientists: One was the community of English-speaking residents, initiated by the Quaker immigrants and their flight from war; the other was the Monteverde Cloud Forest Preserve, which would find its beginnings in the friendship between a conservationist, a chain saw salesman, and a local Costa Rican farmer named Eladio Cruz.

Monteverde, Costa Rica

Under the shade of a tall canopy tree in Campbell's Woods, George Powell laid his bamboo pole and mist net down in the damp leaf litter and stood up to listen. For a few minutes, he heard only bird calls emanating from the boundaries of the forest patch and the open pasture, wind in the treetops, singing water. Then the terrible sound started up again—a distant buzzing that had been gnawing at him all morning. He had been living with it for weeks, and he knew that something would have to be done about it. Powell looked around for his field partner but couldn't see him; he must have gone up the slope to check the nets on the other side of the gully. Leaving his equipment where it lay, Powell struck off through the woods to find the source of the incursion, crossing hillsides and streams and cutting through fields and pastures, going over the old oxcart road that had seen its first jeep only twenty years ago. As he drew nearer to its source, the sound grew louder: the ragged buzz of a chain saw eating wooden trunks, the coughing of its motor, and the crashing of massive trees coming down in the old-growth forest. He was not far at all from the little house that he was fixing up on the Quaker land above the valley dropping down to San Luis. Powell climbed up the ditch to reach the flat land and the rumbling chain saw, and when he did he was met by a holler that sent him running back for cover.

"*Al suelo!*" a cheerful voice called as strong wood snapped and cracked, sending one of the last tall trees on the borders of the farm cascading down into the understory. To be considerate, the chain saw's operator added in English: "*Timber!*"

Birds lit up from the undergrowth when the old tree hit the ground, and other unseen creatures scrambled for safety as sawdust from the cut wood took to the air like smoke. The early-morning sun poured down into the open forest through the gap, and as George Powell stepped into the clearing, Wolf Guindon wiped his brow, set his chain saw down on the body of the fallen tree, and called: "Hello, neighbor!"

Ironically, the first time George Powell and his research partner Bill Buskirk had ever heard of Monteverde had been from Wolf Guindon's cousin Jerry James; and when a professor at the University of Costa Rica also recommended Monteverde, Powell and Buskirk selected the community sight unseen for their research project on Central American birds. They arrived with their wives in April 1970 to find a small Quaker community perched above the local Tico farms, in the shadow of an extraordinary forest that was falling day by day to axe blades and chain saws. For their study site, they selected one of the only large forest

plots that was not actively being stripped for lumber by the Quakers.*
After a year of sleeping in the small cottage and cramped bedroom at
Irma Rockwell's pensión, Powell had approached Wolf Guindon about
moving into a small house on the Guindon property next to Campbell's
Woods, which he and his wife, Harriett, would fix up in exchange for
rent. George would help with the evening milking, and Harriett would
assist Lucky with dinner preparations; the Guindons even invited the
Powells to join their Sunday-evening gatherings.

But while they ate beneath the same roof, the two men saw the
forest through very different eyes. Soon after arriving in Monteverde,
George Powell had discovered that the cloud forest surrounding the
community was a treasure trove of endemic and endangered biodiversity; with every tree that fell to clear a pasture, he felt that
one-of-a-kind ecosystem growing more imperiled—sand running
through an hourglass. He was dismayed by the attitude of some
biologists who had traveled to Monteverde, collected their data and
specimens, and left—giving no real concern to preserving the forests
they had mined for glory. The golden toads were only one example: Jay Savage had published his paper announcing the discovery
of the new-to-science species in 1965; but in 1970 there were still no
formal efforts underway to protect the vulnerable species, and the
world was closing in on them. Mysteries still surrounded the golden
toads (Where did they go when they disappeared after the rains?
How had they evolved to be so bright when their relatives were dull
and unadorned? Did they really only live on the ridgeline of a single
mountain above Monteverde?), and if their highland forests were
felled for farms and pastures, they would take their secrets to their
graves. The little toads had as much a right to live as any others who

---

* The patch of virgin forest was on a farm owned by John and Doris Campbell, two Quaker settlers who would be instrumental in instilling a conservation ethic in the small community. On May 27, 1964, John Campbell took the very first photograph of the golden toads.

had settled on the lower slopes of their green mountain in the recent years. These were the thoughts that haunted George Powell as he lay awake at night in the little cabin on the Guindon family farm. And the golden toads were not his only worry: He feared for the dozens of other species that would vanish with the ruin of Monteverde's forests.

Wolf Guindon, on the other hand, had recently become the first Homelite chain saw dealer for Costa Rica, and he was busy cutting forest to develop a farm in the middle of the golden toad's only known habitat. But Wolf was a seasoned homesteader, and by 1971 it was becoming clear to him that the ridgeline forest was not the perfect site for cattle and agriculture that he had hoped. He and his sons had spent weeks working in the elfin forests along the crest of the Divide, fighting intense wind and rain and wilting under perpetual cloud cover. The strong vines and branches had dulled the blades of their machetes, and they had grown tired to the bone. When the trade winds finally blew them off the mountain and they began the long hike down the Pacific slope toward home, they saw the sun for the first time in three days.

As Wolf passed the windy corner and cut through the bullpen toward the Guindon farm, musing on the wreck of land that he would likely now abandon, he supposed that one person would be satisfied to hear about his troubles: His new neighbor George Powell had taken every opportunity to remind Wolf of his concern for the current rate of forest-clearing in Monteverde—including at their weekly Sunday dinners. Powell feared that he would live to see the disappearance of the resplendent quetzal from the cloud forest, the bare-necked umbrella bird from the wet Atlantic forest in the valley, and the golden toads from the elfin forest on the lonely ridge. When Powell began to seriously discuss protecting the forest, Wolf Guindon thought his friend was in the wrong country.

"There was a need for working farms and pasture and crop production," Wolf remembered later. "It's a very small country and you couldn't expect people to make a living when they couldn't develop the forest for lumber and clear land for farms and buildings."

But some of those forests had desires of their own. The twisted wind-blown woodlands on the ridge were stubborn and hardy, repulsing all of Wolf's attempts to bend them to his will. So, when a local Tico approached him about buying out his claim in the highland forest, Wolf was open to the offer. The local was working on behalf of land speculators in San José, who envisioned establishing a large dairy operation at the top of the mountain. But when Wolf mentioned the offer to his neighbor during one of their weekly dinners, George Powell realized that the situation in Monteverde had reached a tipping point. If that high-elevation elfin forest on the Continental Divide fell to the chain saws and axes that were already at work on the lower slopes of the cordillera, the golden toads would never stand a chance: Their world would end.

And there was another reason that the ridgeline land was worth protecting. When he considered the most vulnerable and threatened forests still standing in the area, Powell had identified three major access trails. The horse path into Peñas Blancas seemed, for the moment, beyond his grasp, because the wild and expansive rainforests of the Caribbean valley were too remote and inaccessible to manage from Monteverde. The Chomogo trail on the Atlantic slope of the Continental Divide ran through an area that Wolf and Powell called "El Valle," where two homesteaders had begun to widen the trail and cut new clearings. The third access point walked the backbone of the mountain through the breeding grounds of the golden toads. Powell knew that if he could buy out the homesteaders and farmers in those key areas, he could slow access to the old-growth forests on the slopes and ridgeline of the cordillera, the chain saws would fall silent, and the wounds would begin to heal.

When Powell revealed his ambitions to his neighbor, Wolf agreed not to sell his highland claim to the developers; he had already begun to turn his attentions to the Peñas Blancas Valley, and his uncle Walter had grown old enough that hiking up into that forest had become a burden. They turned down the offer from San José, and Wolf Guindon and Walter James sold their holding to George Powell for the same price they'd paid for it. With one plot of forest protected and a vision beginning to take shape, Powell asked Wolf for help with another local who had a claim in the ridgeline forests; they would have to reach him before the speculators from San José. Fortunately, many of the local landowners were happy to turn a profit on the undeveloped properties, investing the earnings into their primary farms in the lower elevations. Only the speculators from San José proved difficult to convince: Powell had to call in a favor from two professors at the University of Costa Rica, and their trek to the farm deep in the highlands persuaded the speculators to surrender their machinations to develop Monteverde's forests. In the end, five properties in the ridgeline elfin forest—the only known breeding grounds of the golden toads—would be the first properties purchased for the Monteverde Cloud Forest Preserve.

In the early days, Wolf Guindon's efforts in support of the Reserve were instrumental in cutting trails, contacting landowners, and establishing boundary lines around the areas that he and Powell had identified for protection. At that time, Wolf's motivations were more pragmatic than idealistic.

"I got involved with George in a neighborly way," he recalled, "but the real truth is that I took on the work because of my love of working in the forest, which I was doing for nothing anyway, and the offer of a cash income."

"I believe that Wolf loved the forest and loved to be in it," Powell remembered later. "Before we connected, the only way he knew to justify being in the forest was cutting it down. When he discovered that saving it—being in it to protect it, to cut and measure the growing Reserve's boundaries—was a viable alternative to destroying it, he took to it like a duck to water."

THE GOLDEN AGE AND THE GREEN MOUNTAIN

By 1972, Wolf had brought in a trusted partner to assist in surveying the boundaries of the Reserve properties and patrolling for squatters: a local Tico farmer named Eladio Cruz. Only a few years earlier, Wolf had hired the young farmer from San Luis to help him try to cut a pasture from the highlands on the ridge that he and his uncle would eventually sell to Powell. Now Eladio was walking those same paths in an effort to protect the forest. For the better part of a year, they spent long nights in a refugio* they'd built from abandoned homesteads a two-hour hike from home, taking shelter from the cold and wet and windswept forest around damp firewood that filled their little cabin with smoke. In those long nights on the heels of long days on the trail, elbows knocking into noses while they slept on the cramped floor, their suffering was occasionally allayed by a whisper from the door, an invitation to come out into the cold, wet dark for just a moment—*you won't regret it*—because the golden toads had once again emerged.

Eladio remembered the dazzling species well from his first forays into the dark forests, when he had climbed up the Pacific slope from the San Luis Valley below the mountain.

"In fact, I have known about the golden toads since I was a child," Eladio remembers. "Thirteen or fourteen years old."

In the early settling years, the young Tico had heard about the Quakers that had come down from los Estados† from his older brother when they were working together to develop their father's farm in San Luis, rising early to spend their long days shaping the land, unconcerned with the affairs of outsiders. "My father was a farmer, and he taught us to be farmers also," Eladio remembers. "In those days, it was get up, eat breakfast, get your machete, and go out and plant."

In 1960, when Eladio was twelve years old, his father took him up the trocha and over the far side of the Continental Divide for the first time, across the community of Monteverde and down into the rugged

---

* Refugio: A small shelter or refuge; in this case, a rugged shack.
† Los Estados: The United States.

folds of the Peñas Blancas Valley where the family had a claim to a plot of land deep in the uncharted rainforest. In Monteverde, high on the ridge of the Continental Divide, the moisture that created the cloud forest comes in the form of mist and rain blowing up from the Atlantic slope of the cordillera, and the area known as Peñas Blancas is buried in the deep green tangles of that jungle. It is the domain of mudslides and cabezas de agua* pouring down the Peñas Blancas River, of the deadly terciopelo and the eyelash viper, and of strange trails winding through the dark. Eladio heard folk stories of duendes de la montaña and enanos, disquieting apparitions with small bodies and long fingers said to haunt the forests of the undeveloped valley. The human homesteaders in the area were not any less concerning. As a child, Eladio stumbled into a cacique-fueled fiesta in the jungle, where men fought with machetes for sport, and blood painted the green lianas red.

In spite of the wild and sometimes violent nature of the Peñas Blancas rainforest, Eladio would go on to build his own cabin in the valley by the mid-1960s, when he was working closely with Wolf Guindon as a hired hand farther up the mountain in Monteverde and becoming fast friends with the Quaker settlers. It was from Wolf that Eladio first heard reports of the strange and beautiful golden toads.

"He told me that there was a little toad that was golden," Eladio recalls, "and that it was always there and he would see it almost every year." Eladio would often find the toads in the areas of the highland forest that had been cleared; and as he explored deeper into the undisturbed woodlands—through the swampier areas, or to hidden ponds—their numbers grew. "You could find hundreds," he says, "or thousands, I don't know, in a single day."

As Eladio spent more and more time working in the misty forests, the toads became a familiar sight, and he grew accustomed to their bountiful numbers.

---

* Cabeza de agua: Literally, a "head of water." Cabezas de agua are flash floods known to occur in mountain rivers, unpredictable and deadly.

THE GOLDEN AGE AND THE GREEN MOUNTAIN

"It was always very common to see them," he says, "and so for me it did not have much importance at that time. For me, it was just another little frog—but it did evoke admiration. It's a little toad that I met in the 1960s, and then later I found out that it was so important."

These reports from Wolf Guindon and Eladio Cruz in the elfin forest on the ridge seem to be the earliest recorded accounts of the species; if other modern explorers among the Quaker immigrants or local Costa Rican homesteaders had encountered the toads prior to these individuals, their stories remain unknown.

"Apart from Guindon, I don't know who saw it first," Eladio admits. "It is a pity, because many of the people that were the first to go into that area have passed—and I think they would be the ones to ask."

The golden toads were a welcomed sight, like old friends, when Eladio returned to those ridgeline forests to help Wolf Guindon and George Powell measure the boundary lines of the Cloud Forest Preserve in 1972. George and Harriett Powell had invested their savings into the initial land purchases to secure access to the golden toad's breeding habitat, but they were still a long way off from ensuring the lasting protection of Monteverde's forests. After long days teaching science classes at the local Quaker school, Harriett typed letters on Wolf Guindon's antique typewriter by candlelight, and George sent the word out to his colleagues with connections to conservation organizations, hoping that international support could supply the funding for more land purchases. Over the next few years, their mad scramble to protect the old-growth forests of Monteverde received support from groups like the Audubon Society, the Explorers Club of New York, and the Nature Conservancy. The newly initiated U.S. branch of the World Wildlife Fund was interested in supporting the campaign, willing to make their first donation for land purchasing to Monteverde. Powell, recognizing the potential for Monteverde's most endangered animals to become its icons, drafted a proposal that emphasized the protection of the resplendent quetzal, the bare-necked umbrella bird, the Baird's tapir, and the golden toad.

THE GOLDEN TOAD

While international funding support was necessary for meaningful land purchases, the Reserve would never have succeeded without the involvement of the local people who knew the forests, community, and culture better than anyone else. The spark of a natural conservation ethic that the Quakers had carried with them from the green hills of Alabama was only waiting to be kindled. The local Costa Rican subsistence farmers and homesteaders had left the highland forests above 1500 meters largely undisturbed; so, as the Quakers cleared new elevations for their farms and pastures, the community set aside almost a third of their original land purchase from the Guacimal Land Company to protect the headwaters of the Río Guacimal—the steep, swampy forest sheltering their watershed. Similarly, when John and Doris Campbell established their farm, they preserved more than half their land as virgin forest. John Campbell was also known as Monteverde's unofficial weatherman, mounting a weather station on his property to track rainfall data that would, decades later, prove invaluable to the biologists interested in studying a changing climate at the edge of the world.

The Campbells were not alone. When Walter and Mary James joined the Quaker colony in 1958, Mary taught biology classes in their home and eventually at the Monteverde Friends School; Walter's interest in plants and horticulture led him to collect specimens for the Costa Rican National Museum, supporting the identification of three species previously unknown to science. Their adopted son Jerry, of course, would play his own role in the ecological history of Monteverde with his revelation of the golden toads.

In April 1973, George Powell transferred the 328 hectares of forest in his name to San José's Tropical Science Center, which would manage the newly christened Monteverde Cloud Forest Preserve (locally referred to as "the Reserve" or "la Reserva"). Wolf Guindon signed on as a part-time employee to continue his surveying and patrolling efforts, and to work as an intermediary on local land purchases; the chain saw salesman had proved indispensable to the conservationists. In February 1975, Powell's five-year leave of absence from the University of California at Davis had come to an end, and George and Harriett returned to the United States with three years' worth of funding secured for the Reserve. But to this day, George Powell maintains a cottage on the Guindon farm, a short walk from the protected forest in the hills above the town.

Though unmarked by the scars of human exploration for uncountable years, the ridgeline above Monteverde had become a nexus of crisscrossing identities, experiences, and existences. The Quakers would be tested in their principles and ambitions, their new world eventually swallowed by the next generation of explorers on the green mountain. The golden toads would find their secret world irrevocably revealed, their dark sanctuaries lit up by the spotlight of discovery and renown. And Eladio would return to his father's farm with stories of the treasure he'd encountered, carrying the tales of the golden toads like a secret fire into San Luis and the wilds of the Peñas Blancas Valley—while the mythical enanos listened from the shadows of the bosque, their eyes alight with greed at the news of such a harvest.

With the establishment of the Monteverde Cloud Forest Preserve, the golden toads were safe—for a little while—in their protected habitat. And while they had never failed to inspire fascination in the local farmers, homesteaders, and explorers who encountered them, they were only just beginning to be known to a wider world when the Powells wished them luck and left them to their secrets in the misty woodlands on the distant mountain. That would all change over the course of the next decade, when the golden toad would rise to the pinnacle of its fame and glory—a few years in the light of splendor before its last magic trick: vanishing into the dark.

Monteverde, Costa Rica

Mills Tandy had traveled to Monteverde with an Exakta IIa camera and a few rolls of Kodachrome 25 35mm film. In the hours between

THE GOLDEN TOAD

4:00 a.m. and sunrise on Friday, May 24, 1985, the biologist's empty tent fluttered in the soft night wind moving through the elfin forest as he haunted the woodlands with his flashlight, his field notebook, and his camera. For the last four days he'd walked these trails every few hours, beneath the tall trees and through the gnarled dwarf forests that guarded the treasures on the ridge. While his empty tent sat abandoned in the soft, wet earth, the biologist was crouched beside a shallow pool of water protected by thick buttress roots, watching the golden toads.

Among the dark brown roots, the green leaves textured with epiphytes, and the black soil like bone showing through the leaf litter, many of the little orange toads were clambering over one another in the shallow water, searching for the darker females and trilling short release calls when clasped or disturbed by one another. Others sat like stoics along the boundaries of the breeding pool, golden statues perched on wet twigs and soft moss. In nearby pools, there were strings of eggs laid in dark pockets beneath the tree roots, where small brown tadpoles were just beginning to come alive. Reflections of the golden toads danced in the water.

Around 5:00 a.m., the sun began to come up east of the cordillera, and the biologist left the little pool to wander back to his campsite in the early-morning light. He took off his rubber boots and started the coffee brewing, then made a few notes about his observations and laid the camera beyond the reach of creeping water. He set an alarm to wake him up in thirty minutes; soon he would stir from sleep to warm coffee and a wet forest hopping with golden toads.

The first golden toad that Mills Tandy ever saw was in a genetics lab in Austin, Texas—a long way from his little tent on the green mountain. A group of researchers from the University of Texas had collected several of the bright orange males to use in hybridization experiments while Tandy was in the lab for his graduate work on African toads. He had never seen anything like them. When a colleague later offered him the opportunity to purchase a piece of land in Monteverde, his memory of those strange toads lured him to the tropics.

As he sat up through the long nights on the ridge, he tinkered with his camera by the light of a small lantern, loading film and cleaning lenses to ensure that everything was working properly. He had seen a few photographs of the golden toads, but he also knew that many of those images had been taken at an artificial site: a pool of water in an excavated pit created by researchers studying cloud-forest soils. When Tandy had started planning his own study, he'd begun to wonder if he could photograph the golden toads himself: in undisturbed elfin-forest habitat in natural light.

With help from Wolf Guindon and Giovanni Bello—a local Costa Rican forestry graduate and codirector of the Reserve who had been identifying the active breeding pools—Tandy had selected eight study sites along a one-kilometer section of the breeding grounds with the intention of surveying the toads' behavior night and day for two weeks. Wolf would come up once or twice to visit, and a local field assistant would carry up a cache of supplies the second week; aside from that, he was alone.

"I didn't see anybody else," he recalled later. "Just the toads."

At the height of the breeding activity, he counted 164 toads in one day.[*] The site with the highest aggregations was a pool of calm water beneath the roots of a dwarf umbrella tree. The roots had grown around the remains of a fallen tree, and the tree had decayed, leaving an empty bowl where the water was protected from runoff and heavy rain; the golden toads had made their place of ritual in the space left by a ghost. When the breeding season came to an end, Tandy packed up his campsite and hiked down the mountain, leaving the reclusive species to withdraw again beneath the earth, until the new rains called them back to breathe fresh air on the high mountain. Not long after Tandy's return,

---

[*] These were separate observations, but some were likely repeated observations of the same individual.

the biologist and photographer Michael Fogden sought him out to ask him what he had learned from his two weeks camped out in the elfin forest: In Monteverde in the 1980s, knowledge of the resplendent species was better currency than gold.

When Michael and Patricia Fogden arrived in Monteverde in 1979, they both came with doctorates in zoology and a long history as research biologists in Africa, North America, and the United Kingdom. Looking to begin a new career as freelance naturalist photographers and writers, they quickly acquainted themselves with the unique and endemic treasures of Monteverde's cloud forests—including the golden toads. It would be the Fogdens' images that brought the golden toads to the world—or the world to the golden toads. Many of those photographs were published in their 1984 *Natural History* feature, "All That Glitters May Be Toads," which explored the habitat and mating behavior of the secretive species. "To reach the area where the toads live," they wrote, "it is necessary to climb narrow, slippery trails, ankle-deep in mud. On the exposed crest of the cordillera one enters an elfin forest of gnarled, stunted trees covered with spiky plants, orchids, and thick carpets of moss, where drifting mist deposits a film of moisture on every leaf."

The Fogdens' photographs also introduced the world to the explosive breeding habits of the golden toads. The cover image of their story featured more than two dozen bright male toads in a small, dark pool: water black, leaves green, toads brilliantly orange. Many of the males sat perched on the edges of the pool; some swam beneath the surface of the dark water; one pair was in amplexus*; others seemed to levitate on the surface of the pool itself, reflections warped and muddied in its mirror. Other images showed floating eggs like planets, spheres of life in the water's cradle. "Suddenly the ground is full of golden toads," they wrote, "vivid splashes of color around a small forest pool. They sit so

---

* Amplexus: The mating position of frogs and toads in which the male clasps a female with his front legs, riding on her back in a position to fertilize her eggs.

still and are so brilliantly colored that they seem more like diminutive gold-painted statues than living animals."

It didn't take long for the Fogden photographs to become ubiquitous in the small community of Monteverde; images of the bright males and spectacled females adorned postcards and advertisements from lower Santa Elena up to the gates of the Reserve. But the true impact of their photos was just beginning, and would reach far beyond the slopes of the green mountain. Ecotourists were beginning to follow the footsteps of the biologists up into the mists of Monteverde, and it was impossible for the new visitors to the mountain town to walk along the sidewalks of the central square without passing under the gaze of the golden toads. By 1981, the old oxcart road up to Monteverde had been graveled, and the popularity of the Cloud Forest Preserve was growing—as was the celebrity of one of its most famous residents. A 1981 *National Geographic* article was one of the first pieces of media in the United States to include a photo of a golden toad. "Savior of a primeval forest," read the caption, "the golden toad . . . helped bring about the creation of the Monteverde Cloud Forest Reserve." The growing tourism industry was helping to protect the forests that had once been at the mercy of axes and chain saws. "Costa Rica," *National Geographic* reported, "with thirteen new national parks and six biological reserves, leads Central America in moving away from the soil-leaching deforestation that has long plagued the isthmus."

As mounting crowds of biologists and tourists made the pilgrimage to Monteverde to see the famous species for themselves, others looked for ways to bring the golden toads to the wider world. The Metro Toronto Zoo in southern Canada contacted the Tropical Science Center in San José to request three pairs of golden toads for captive breeding; the Tropical Science Center denied the request. For a few years in the late seventies, the Reserve kept a few toads in a terrarium at the casona* as a learning exhibit, and to give visiting tourists a chance to see

---

* Casona: Literally a big house or mansion, the casona is the colloquial name for the visitor's center at the Reserve.

Monteverde's famous golden toads without hiking up into the fragile forests. Toward the end of each season, the captive toads were released back into the elfin forest and replaced with new specimens—small deposits and withdrawals from an apparently limitless reserve.

Limitless though the toads might appear, the staff at the Reserve considered it their responsibility to watch over the golden toads—one of the species that the Cloud Forest Preserve had been created to protect. The Metro Zoo's request to acquire a few pairs of the iconic species had come through the appropriate bureaucratic channels, but there were reports of other incidents that did not. Irma Rockwell once called the Reserve to tell Wolf Guindon she had seen a little girl carrying a golden toad down the road past her pensión; the girl's parents had let her take it from the breeding pools when she had cried to carry one of the beautiful orange creatures home. Mills Tandy was at a conference in Bonn when a German pet dealer tracked him down to ask if Tandy could help him acquire golden toads, and Wolf Guindon heard about advertisements in trade papers in the United States as well. "Apparently, you actually could buy one for five dollars," he would remember later. With no poachers or private collectors ever caught red-handed (besides the little girl who made the poor decision to stroll past Irma's pensión, holding glory in her hands), it is impossible to know how many golden toads were ever carried out of Costa Rica. It is not hard to imagine the gleaming delegates of a foreign country in the dim light of a basement, caught between glass walls and lit by an orange light as brilliant as bright toads in black pools far from home.

Excluding those unlucky few smuggled into the back rooms of German pet shops (and the specimens preserved in formaldehyde in the world museums), the golden toads lived and died in Monteverde, that great cerro* like a well of creation.

---

* Cerro: A hill or mountain—usually used to refer to a specific mountaintop (Cerro Amigos, for example).

THE GOLDEN AGE AND THE GREEN MOUNTAIN

It might be true that these years saw the golden toads at the height of their acclaim, when their renown had worked its magic on their mountain, shaping its future. They inspired the protection of the cloud forests and drew foreign biologists deep into the highlands to mark the boundaries of the new reserve; as their images traveled across oceans, they would leave their mark on the next generation of explorers too. The foundations of Monteverde were beginning to shift from an agricultural economy to ecotourism, and the locals who had once pulled down tall trees to farm the land were beginning to make their living by simply allowing those trees to stand.

Those years might have been the height of the golden toads' acclaim, but it is unlikely that they were the height of their existence. We can wonder if their lives were improved by people's knowledge of them, but it is a question that might transcend the way we think about ourselves. They had lived for untold time without any understanding or awareness from our species, never missing our esteem or adoration. When they had first emerged into the long, dark, and quiet midnights, the dim sparks of our young fires did not yet flicker in the valleys far below. They had watched their generations live and die as they handed down their precious secrets; through dry spells and heavy rains, they clung to their sacred rock, the temple of their existence. In those long years of solitude, the roots of twisted trees clutched to find the soft pockets of the earth, and their hollows collected rain pools beneath a dark sky lit by purple electricity. That was the promised land of the golden toads: raw and ragged on the fringes of creation. There was no want for surplus, no fear of god; only the amber pools of light burning like torches in the dark—the first golden toads living and dying on their elder mountain. That was the golden age for the golden toads.

But a golden age can never last. The ridgeline elfin forest above Monteverde was like a distant throne for a kingly species, but it could not raise the golden toads beyond the reach of something dark and deadly that was creeping softly up the mountain.

THE GOLDEN TOAD

...

The short scene offered in the following pages will be the last before we must begin to say goodbye to the golden toad. It is not the introduction of its killer (though that meeting will be coming soon, I am sad to say), but only a reminder of the ways that people loved it, looked out for it, did what they could to protect it while it lived. If you had been there, I imagine you would have done the same.

One dark evening, Wolf Guindon fought the wind going up the road to the Reserve. The rain had turned to mist at sundown, but in the night the moisture brought a chill, and he wrapped himself in his coat and pulled the brim of his hat down as he passed into the shadow of the casona, leaving the silver light of the moon. He conjured a ring of keys from his pocket and began to search by touch for the one that would open the lock on the door before him. When he found it, he slipped into the casona and let a seam of golden light spill out through the small gap in the open door, like a long finger pointing out into the night. For a few minutes, the evening was still and the scene was silent—the canopy rustled as a night-creature moved from one branch to another; a lonely call emitted from the deep green forest; a lullaby of running water wandered up from the quebrada. Then the golden light from the open door expanded, like the sun rising in a flash, and disappeared. A key turned in a lock once more and Wolf fled back down the dirt road with the wind hollering behind him. Underneath his coat, he carried a small jar with living treasure trapped inside.

He moved between the rising walls of forest like a wraith, striding surely in the dark to the corner where the road dropped down to the trocha, where he broke off and climbed the bank and cut cross-country beneath the tall trees that stood guard around the pastures. The cows watched him from the edges of the open fields as he struck off for a little house on the far edge of the forest.

THE GOLDEN AGE AND THE GREEN MOUNTAIN

Stepping into the light, Wolf shook the water from his coat and hung it up to dry, moving to the kitchen table and opening the little jar. He transferred the two passengers into a different container, and then stood watching the pair of golden toads, evacuees, settle in to their new abode.

For the last few seasons, the toads in the terrarium at the Reserve had been dying. Wolf had mentioned the trend to a few biologists, but none of them had taken up the concern to study the problem. Afraid that there was something wrong with the terrarium itself, he'd decided to bring the captive pair down to the Guindon farm in hopes that they would fare better in a different vessel.

As Wolf watched them through the glass, he was hopeful that the two bright-orange toads would survive, maybe even long enough to be released back in the breeding grounds, set free, exchanged for others who had lived for years already on the elysian mountain. But in the morning Wolf would find that the golden toads had expired, two more victims of a calamity that he did not yet understand.

For there was another traveler stealing through the secret trails. This one was a pilgrim, too, old and hungry and dealing death; and as the curtain fell on the golden age of the green mountain, its hour had finally arrived.

CHAPTER THREE
THE END OF THE SHOW

San José, Costa Rica

The ghosts were restless on the streets of San José. Some dozed politely in crumbling tumbas* and family vaults in the mossy cemeteries, but others had begun to stir again as their names were spoken and their graves disturbed. As my fiancée, Pri, and I made our way across the city, the disembodied voices of old women shouting lotería numbers echoed from the alleyways, and we passed beneath the shadow of the National Museum where old bullet holes from the country's civil war flecked the yellow paint like the fingerprints of the dead. Dark clouds above the encircling mountains prophesied approaching rain. We were walking fast because we had an errand to accomplish in the city before catching our two o'clock bus—a ride that would carry us along the salty shipping coast of Puntarenas and up into the old hills of the green mountain, where the ghost we sought was waiting somewhere in the hidden breeding grounds of the golden toads.

It was a restless spirit that had drawn me back into the tropics. In the nine months I'd spent in the United States after leaving Costa Rica, I'd been unable to shake the image of the "Lost Frogs: Most Wanted" poster I'd discovered in the study center, and the memories of my father's stories it had reawakened. I'd spent that winter reading everything I could find about the golden toads—scientific papers, travelogues, old pages of *National Geographic* exhumed from my university's archives—but the accounts were often inconsistent, vague, or incomplete. Reports varied

---

* Tumbas: Tombs.

on the species' range, on where they went when they disappeared after the rains, on who had seen them last, and when, and where. When Pri joined me in the United States in January, as snow fell on the Northern Arizona pine forests around our small apartment and the cold nights turned her tropical blood to ice, I mined her memories of her biology lectures at the University of Costa Rica. She told me all the stories she remembered, but they didn't cure my fascination—didn't break the spell, didn't exorcise the spirit. I had a growing list of questions that hadn't been answered in the fifty years since Jay Savage's first encounter with the golden toads and the thirty years since their last appearance, and I had begun to suspect that the truth was not waiting in the books or magazines or scientific journals: The truth was hidden in the green hills of Monteverde—the only place it had ever been.

I was working my way through a graduate program in documentary studies at Northern Arizona University, and I owed my advisers a thesis project in a few months. I'd heard that there was funding available for international research, so I persuaded Kyle to help me sketch a story structure, and I pitched a documentary film about the golden toad. *It will be a journey into the past*, I wrote, *and a return to the hidden breeding grounds of the iconic species*. I made optimistic reservations at the university's gear room: a tripod, a mic kit, a Canon 60D, and a waterproof bag. My thesis committee signed off on the proposal, and in February—just a few weeks after Pri moved to Arizona—I received word that my request for funding had been approved. When the snow melted, I would be going back into the tropics.

Pri had barely unpacked in our little studio apartment when we started planning our return to Costa Rica. While I descended into the archives to look for clues to the beginning of the trail, Pri began to organize the logistics of our efforts. She called old friends at the Reserve to appeal for access to the restricted trails, and emailed old classmates and professors at the University of Costa Rica who had spent years studying the amphibians of the tropics. It had been a few years since she'd worked professionally as a tropical biologist, but she knew far more than I did about the natural history of the cloud forest. She stayed up late answering my questions about the scientific papers I was reading and envisioning our expedition through the mythic forests. I was grateful that she would be beside me on the dark paths of this long and uncertain journey. The last time I'd set off into the tropics, I'd left everyone I'd known behind; this time, I was going back with family.

When we touched down and set off through the streets of San José in May 2018, I had to jog to keep from losing sight of Pri as she laced her way among the crowds of people on the central avenue; her city instincts had not deserted her at all.

"Come on!" she shouted back to me; she was still buzzing from her reunion with Costa Rican coffee. "They're closing soon!"

From what I could discern, the exact location of the breeding grounds was a closely guarded secret, missing from the old reports and scientific papers we had found. To locate the trail that wound into the haunted forests of the golden toads, we would need a map, a name, and a guide. In San José, we were closing in on the first of the three. That morning, a dour government employee had sent us across the city to find the Civil Registry where they still kept old maps of the places we were hoping to explore and film, and they would sell them to us—or so the frowning woman at the Municipality had said. The historic breeding grounds of the golden toads had been closed to visitors for decades, and the old trail had been erased from the maps that the Reserve handed out to tourists. In my research, I had found shadows of the path that snaked its way along the ridge and among the hidden pools, but I didn't know its name, and I didn't know where it began. We needed a map—an old map—to trace the elevations we'd discovered on the field notes of the first collectors, to find the entrance to the lost world of the golden toads.

When we reached the Civil Registry, we found the geographic department buried belowground like a cellar, and we descended into a musky darkness that felt cavernous. In a dim glow at the bottom of the stairs, an old man in spectacles sat hunched behind a desk surrounded by huge maps like wallpaper. When he noticed us, he glanced up from his work with the look of a hermit sitting down to dinner and not expecting company. While Pri gave the clerk the names of the transects* we were seeking, I studied the topographies that lined the walls, renderings of Costa Rica's most secluded cerros, volcanos, and coastal rivers that reached with long fingers to the sea. Finally, the bent man stood and shuffled over to a cabinet with deep drawers, where he sifted through reams of wide papers and withdrew a pair of maps, laying them down on the desk as dust rose in the strange light. He sniffed and set a finger down on the thick paper and announced, "Aquí está: Monte-Verde."

The maps told the elevations of the Tilarán cordillera's undulating ridgeline above the mountain town and noted small streams like blue veins running down the hillsides. I had recorded the elevations from Jay Savage's field notes, and now I could put my finger on those certain places on the maps. These were the most detailed surveys that we had ever seen, revealing indirectly the corridors that the golden toads had once inhabited, and the ways that one might travel to rediscover those abandoned havens. With the maps rolled up and tucked beneath our arms, we thanked the old man, left a small tower of colones on the desk, and climbed back up into the light.

A message from my brother Kyle appeared on the dark screen of my phone: *Did you find the maps? Do they show the way?* He was three thousand miles away in the Everglades right now, looking for endangered panthers in the flooded forests; but in recent days his dispatches had been replaced by questions about our odyssey. I read his message and put my phone away; there was no time to answer now.

We hailed a taxi and sped off across the city toward Terminal 7, where the public bus was idling, waiting for us to climb aboard for the

---

* Transect: A scientific term referring to lines that cut through natural landscapes, enabling standard measurements.

long drive up the mountain. As afternoon gave way to evening, the big bus carried us up the winding roads, from San José along the ragged coast of Puntarenas, then up through dry Guacimal to the foothills of the cloud forest; the same roads that Jay Savage and Norman Scott had struggled up when they were little more than ruts for the oxcarts in greedy mud; the same roads that my parents had followed thirty-one years before when their old bus bounced along the crumbling edges of steep canyons; and the same roads that I had taken on my first excursion into the tropics, now almost three years in the past. As we drew closer to Monteverde, we passed the billboards advertising night walks and coffee tours, bungee jumping and hanging bridges; the small community, once a farming town, had continued to evolve. But the Cloud Forest Preserve was still standing; the forests that had sheltered the golden toads lived on.

To find the hidden breeding grounds of the long-lost golden toads, we would still need the name of the forgotten trail that would draw us into the tangles of history and a guide to lead us through the beguiling forests. But the map that we had bartered from the Civil Registry was spread out on my lap in the weak light, and I traced the cordillera with my finger a dozen times across the long and restless ride up from the Central Valley. For the first time since I had seen its glowing visage flickering on a screen, the golden toad felt close, as if I might reach out and touch it.

An eerie mist was blowing through the low sun's fading light when Pri and I reached Santa Elena. The little shops were closing for the evening and the patios and bars were coming alive; disembodied forms glittered in hazy windows. When the old bus rolled into the Centro Commercial* in Santa Elena and pulled up, purring, at the foot of the steep hill, we piled out with the other passengers and watched them pour like army ants down to the hostels in the square. A few Tico

---

* Centro Commercial: The central "mall" in Santa Elena, a collection of restaurants, shops, and grocery stores.

THE GOLDEN TOAD

families lingered in the parking lot with their umbrellas open to the mist, waiting for their rides to pick them up and carry them off to more distant neighborhoods—Las Nubes or Los Llanos,* where the family farms still ran in relative solitude, not yet sold off as adventure parks for the tourists. On the edge of the shimmering pavement, an old dog waited for his turn to cross the street—watching as the bigger cars and trucks went past, comprehending his place in the universe. We pulled our bags and camera gear from the bus's rumbling guts and walked out to the road leading up to higher elevations.

The local bus was finished for the day, and it was a few hours before the taxi piratas† would begin to line the sidewalks outside Bar Amigos and Taberna in the square, their interiors thick with cigarette smoke and their mufflers sending hot exhaust into the night. So Pri and I heaved our duffel bags onto our shoulders and walked through town on foot. On the broad side of the shopping center, the Roberto Wesson mural arrayed a bright menagerie: iridescent motmots and resplendent quetzals perched on woody limbs above foraging agoutis, and a slinking ocelot with glowing eyes peered around the corner of the painted wall. A little farther up the hill, we passed the old ranario,‡ where speckled mountain toads and crowned tree frogs peeked out from soil and leaves through the wet walls of terrariums for the tourists to contemplate. Inside, there was a tall glass box standing upright, empty in the memory of the golden toad, a monolith in the dark.

At the top of the hill, we could turn and look down on Santa Elena, a patchwork of small homes and restaurants and gift shops piled atop one another like stone islands in the encircling forests and rolling hills. We walked along the shoulder through Cerro Plano, past little sodas§

---

* Las Nubes, Los Llanos: More rural neighborhoods a little outside Monteverde / Santa Elena.

† Taxi piratas: Literally, "pirate taxis." Unofficial and independent taxis that operate after midnight.

‡ Ranario: Frog pond; an amphibian zoo in Santa Elena.

§ Soda: A small open-air restaurant.

and sleepy markets with their doors and windows open, letting in the last light of the evening. A gang of roaming dogs led us past scattered neighborhoods where empty hammocks hung from tin roofs; a cat on a windowsill offered a cursory hiss, and the pack turned off the main road, heading down a steep driveway to find a porch to sleep on for the night.

Before long, we found ourselves at the foot of the abandoned Sapo Dorado Hotel. The eager jungle was creeping in, but the stained-glass windows were still intact; the solemn images of golden toads burnished in the glass looked down on us like martyred saints. The sun slipped below the mountains and shadows consumed the hollow lobby of the Sapo. We didn't look back as we followed the winding road around the bend and up into a cold, damp wind.

The mountain town was sleeping by the time we reached Monteverde proper in the dark; its golden lights were curtained, and the first pinpricks of distant starlight lit the sky through a scattered veil of canopy. Green arms of forest rose up on either side of the dirt road, and from out of the wet dark drifted the chirps and trills of night frogs singing from a forest of bromeliads, scattered like rain. The orange streetlights, hazy in the mist, were riven by long belts of shadow; here and there in the woods, a dim glow from one of the old houses winked through the trees. Dark and bent forms scuttled across telephone wires.

We could hear the plodding of a slow parade of cows coming down the road ahead of us, out from one of the pastures near the Quaker lands above the trocha. Six or seven of them wandered through, filling the street, and we put our heavy bags down for a few minutes to watch them: aimless, stoic, slouching toward green and dewy beds. The other night-time sounds fell away as an old campesino on horseback trailed behind them, trotted past us, and was gone. Then there was just the white moon peering through the forest like a wide eye, and the trickle of the quebrada running down stony banks in the dark. The sea of forest was all around us, and I knew that soon we would venture in, though I was not certain what we would find waiting for us.

We left the stream, left the road, and followed a thin track up a steep hill; before long, the shy frogs had taken up their symphonies again.

Monteverde, Costa Rica

From her open window, Martha Crump watched the storm break on the mountain. It had been raining for two days; the wind was howling like a lost dog, and old trees groaned as the storm crashed against the dark forest. The night was filled with glowing eyes and muffled footsteps on soft leaves in the rain, a changing of the guard as the wet season and the witching hour settled over the highlands together. It was April 1987, and Crump had been in Monteverde since March observing harlequin frogs along the banks of the Río Lagarto, but the water levels had been low and the moss gray and dusty on the rocks. Now the green season had descended over Monteverde at last, and the rains would quench the thirsty riverbeds; she just hoped they didn't wash the frogs away as well.

She had put the children to bed and cleaned the dishes and her husband had disappeared into the bedroom, sorting equipment for the next day's fieldwork. Crump had just begun to prepare her own pack for the morning when she heard the unexpected sound of rain boots on the wooden porch. When she went out to investigate, peering into the dark, she found Wolf Guindon drenched from head to toe and beaming at her in the stormy night.

"Marty," he exclaimed, "the golden toads are out!"

Images of the famous species flashed in Crump's imagination. When Jay Savage had published his discovery of the enigmatic species in 1965, their magnificent appearance, peculiar habits, and narrow ridgeline of existence had flowed through the scientific community like an electric current. Twenty years after their discovery, there were plenty of questions about the golden toads that still remained unanswered. Crump had not traveled to Monteverde to study them; but on the evenings of her long days in the field, the golden toads were often in her thoughts.

THE END OF THE SHOW

Wolf's excitement was infectious; but in the face of the pounding rain and the beating wind, there was little hope of finding the way up to the breeding grounds in the night: The path was overgrown and the ridge was steep and treacherous. Wolf proposed that they meet the next morning at the Reserve and hike up with the sun to the enchanted ridge where Wolf had seen the toads emerging, like living lanterns in the dark.

Wolf made one last promise to Crump that night before disappearing again into the storm: "Marty, you're not going to believe it!" His familiar howl lingered on for a few moments, filling the air around the house, and then it, too, was gone. The night was left with the lullabies of dink frogs and the rain on elephant leaves as Crump turned the porch light out and retired for a restless night. Far above on the distant ridge, although there was no one there to see it, the stunted forest was alive with golden embers so close to heaven they could almost touch it: a theater of secret fire burning in the beating storm, nestled on a mountain peak that scraped the stars.

In the blue light of early morning, Marty Crump met Wolf to walk together up the feral trail into the windswept highlands. Cold and soaked to the bone, they made their way through deep mud and across slick roots into dark woods. Over carpet dense with moss and leaf litter, they climbed higher into clouds, until they reached the long backbone of the Continental Divide. They passed out of cloud forest and into elfin forest. High on the ridge, the trees grew mangled from the wind, and the world was wet, and climbing epiphytes swallowed nearly every earthly body.

The pair emerged around a bend, and the sight ahead stopped Crump cold in her tracks. Later, she would describe the toads as "jewels scattered about the dim understory," but in those first moments she was speechless. She stood and watched the golden toads as they gathered in the shallow rain pools that had collected overnight among the twisting roots and fallen limbs, defending their stations and searching for mates among the muddy chaos of their world. The only sounds were the howling wind and the rustling leaves among the falling rain, and the music of Wolf laughing happily beside her, reunited once again with his old and faithful friends.

They spent that morning on the ridgeline watching the toads. The shimmering orange males often wrestled over one another and moved through root and ravel to reach the nearest female, tension sharp and passion buzzing in the air. Crump spent hours moving back and forth from one pool to another, making notes and taking photographs to document the rituals of explosive breeding. Wolf took in the sight before him as one might watch a sunset: quietly reverent, confident in its immortality.

As the day grew late, they took one last look at the golden toads before beginning their long descent, leaving behind the pastoral scene of old eternity, a warm glow filling the dark. As they made their way back down the mountain, Crump determined that it would be impossible for her to ignore the golden toads and their strange ethereal world as she pursued her other research. She told Wolf that she planned to hike up to observe the toads every day until they disappeared again, and he promised to join her when he could to assist in her fieldwork. She was mapping out one of the most ambitious scientific studies of the species' breeding behavior, and the two parted ways that night humming with anticipation: What new revelations might they discover among the secret meetings of the golden toads?

In Wolf Guindon, Marty Crump had the best field assistant she could have asked for. As he'd explored the Monteverde highlands to help George Powell draw the boundaries of the Reserve, Wolf had also begun to identify the primary breeding areas of the golden toads. In 1972, Wolf had volunteered to assist the fieldwork of a young biologist named John Vandenberg, who had traveled to Monteverde to compare the growth cycle in metamorphosis of the golden toad to Holdridge's toad (*Bufo holdridgei*), a species that lived in a similar habitat on the slopes of the Poás volcano in central Costa Rica. While Vandenberg scaled the slopes of the volcano, Wolf searched for undocumented breeding pools in other areas around Monteverde, beyond the windswept elfin forest on the exposed backbone of the cordillera. On a long walk toward the Arenal volcano northeast of Monteverde, where the streams drained toward the Caribbean, he was crossing through the primary forests in the El Valle sector when he discovered several pools with a small population of golden toads. Wolf was surprised to encounter his little orange friends so far afield, and in an ecosystem so different from the breeding grounds in his once-imagined pasture. "It wasn't elfin woods or an exposed ridge," he noted, "and had more rain and less wind." Wolf and other guardaparques* at the Reserve began monitoring the El Valle site as well.

Vandenberg's 1972 study didn't pan out in the end—the variability of the golden toads' emergence proved maddening, and he would often receive news from Wolf that they had finally appeared when Vandenberg was far away on the steep slopes of Poás digging for Holdridge's toads—but the revelation that the golden toads had colonized an area outside the boundaries of the ridgeline breeding grounds was like a glimpse behind a curtain hiding a larger world. For the last ten years, general knowledge had held that the narrow, windswept ridgeline above Monteverde was the only place in the world where the golden toads occurred. The discovery of a second site, close in distance but so different from the habitat at the top of the mountain, cast doubt on the assumptions that had been made about the hermitic species.

In 1977, Vandenberg returned to undertake a more dedicated study of the golden toads. As a localized thunderstorm marked the onset of breeding activity, he conducted thirty-five surveys of the primary habitat that spring, making detailed observations during the mating bout. Through the course of his study, Vandenberg determined that the geographic and altitudinal range of the golden toads was extremely narrow, "apparently restricted to elevations from 1480-1600m over less

---

* Guardaparques: Forest rangers.

than a 10km^2 area." Using toe clips to mark the toads, he counted as many as nine hundred males and eighty-eight females in a single day. The day before, he'd observed eggs for the first time—large eggs in small clutches—deposited in two pools filled by the torrential rainfall. The breeding pools were mostly temporary caches among tree roots or depressions in the soil; on one occasion he observed eggs deposited in a pool of water that had collected in a bootprint.

Breeding itself was a fraught process. Roving "balls" of toads formed when up to ten males latched on to one another to grapple for position, competing for mates among the roots and hollows. But the most fervent attacks were directed at mating pairs: Over one two-hour observation period, twenty-three males attempted to displace a male in amplexus, but none was successful. These efforts continued after dark, until there were only a few males left in the pools, desperate to reproduce and missing their chance. Only fourteen out of every hundred males obtained a mate.

Vandenberg described two distinct calls of the male toads: The first was a release call performed when they were disturbed by other males (or by Vandenberg himself): "a low-intensity trill, often accompanied by body vibrations." The second call was less frequent, a soft "tep-tep-tep" like wooden spoons clicking together, from solo males perched in upright positions or in amplexus. Vandenberg theorized that these calls might be associated with courtship or breeding, but that visual cues were more likely to influence mating than the calls that he'd observed.

Through the end of April and into May, the toads appeared sporadically in response to rainfall, but the curtain seemed to have fallen for the season on their perennial show: One by one, they descended to their hidden sanctums to await the next year's rains. Vandenberg collected eggs from eighteen pairs of toads, recorded final readings for the air temperature and rainfall data, and descended down the mountain at the end of May. For everything he had learned about the novel species, he had twice as many questions. "How far golden toads travel to reach their breeding sites is unknown," he would write later. "They are infrequently encountered during other seasons of the year." He suspected that they were fossorial, living underground for long periods* like Holdridge's toad, because he had excavated two adult males and several juveniles (browner with pale blue-white spots, and black and white mottled bellies) after shallow digging. But other secrets of the subterranean toads, for the time being, remained unknowable. More than a decade later, Vandenberg would combine his 1977 data with notes from another biologist's 1982 observations, writing: "No frogs marked in 1977 were observed during the 1982 study."† Other species in the same genus were believed to have a lifespan of around ten years; it was strange that the golden toads were not surviving long enough to be counted twice.

In 1979, Wolf returned to the El Valle site outside the primary breeding area to find a small spring-fed pool filled with the most eggs he had ever seen—close to ten thousand, he would estimate later. There were no mature toads to be seen, and the small pool was drying up, leaving the eggs dehydrated and desiccated. Wolf collected as many eggs as he could carry—almost three-quarters of the whole—and hiked them back to the field station, hoping he could hatch the young toads in captivity and then return them to the pool. But none of the eggs that he collected ever hatched, and when he returned to El Valle later on that season, the pool had vanished.

If the 1970s were the renaissance of the golden toad, the 1980s were the beginning of the end. Marty Crump began her 1987 season with full knowledge that she was not the first biologist to climb into the misty highlands to observe the golden toads, but she didn't know that her study would be the last. Her fieldwork began as routinely and optimistically as any of the others that had come before her. Beginning April 8,

---

* Some species of toads (like the Spadefoot toad of North America) are believed to spend long periods of time underground to survive in dry environments or evade predators.
† The biologist Susan Jacobson published her observations of the golden toads in the winter 1983 issue of *International Wildlife*, under the title "Short Season of the Golden Toad."

THE GOLDEN TOAD

Marty Crump and Wolf Guindon spent two weeks hiking the twisting trail through strong winds and mist, and each day they encountered hundreds of golden toads, crawling out from underneath the root hollows and leaf litter to gather in the wet depressions that collected at the base of the stunted trees in the elfin woods. There, the toads would congregate to worship the precious gem of water, fall in love among the muddy roots, and disappear again.

Crump made observations on their mating behaviors, measuring the size of every toad and recording the ways the males competed for affection from the solitary females. She spent long hours hunkered on the muddy ground, blending in with mist and mosses, writing down the story of their home.

Toward the end of April, Crump was climbing the long and dark trails early in the mornings, and as the toads dipped back one by one into their subterranean retreats, she made a disconcerting observation: With every day, the breeding grounds were growing drier.

"'86–'87 was an El Niño year," she would note later—an irregular pattern of ocean warming that carries unseasonably dry conditions across the face of Central America. "And there was hardly any successful reproduction." Root hollows that had once held brimming rain-catches were dry as bones; leaves crunched underfoot, and there was no sound of water—no pools among the rock. Crump found dried and desiccated eggs in twisted roots where the rain and dew might have collected for a few hours before evaporating. Tucked away with the last faith of their forerunners, the young held on as long as they could; eventually, their shallow pools vanished.

Heavy mist rolled in in early May, and the golden toads reemerged briefly for a second breeding bout when the long-awaited rains finally returned to the mountain. But tadpoles did not appear in any of the pools that Crump encountered; none of the offspring from April had survived. New eggs glimmered in the waters at the end of May, and the toads disappeared again. "If the mist and drizzle continue this time,"

Crump recorded hopefully in her field notes, "the tadpoles in their forest pools might develop and transform."

But when Crump left the mountain to travel back to her home in the United States in July, her final field entries for the 1987 season noted that the few remaining tadpoles she had witnessed on the mountain would not be sufficient to replace the adults that would die over the coming months. She had counted only twenty-nine tadpoles in the evaporating pools: "The toads will need better luck next year."

In 1988, the spring was drier than Wolf Guindon had ever seen it. The high elfin forests that were perennially cloaked in mist and moisture collapsed into a thirsty landscape, scarred by a broken promise. He had been going out early that year, trekking up to the breeding site for months, but he had found no sign of the old familiar congregations. Eventually the rains arrived, watering the dry bones of the mountain, but the golden toads did not emerge. In early May, Marty Crump returned and hiked up to the breeding pools daily, willing each bend in the crooked trail to reveal a triumphant homecoming, hope growing heavier with each endeavor. For weeks, she waited for them, while Wolf crossed miles, up and down the steep slopes of the Continental Divide, searching for his old amigos. He counted nine toads in an area five kilometers southeast of the ventana,* but the primary breeding grounds on the high ridge of the mountain were dark, and the golden sparks that had lit the understory so beautifully had disappeared.

Finally, at the end of May, Crump was pushing through the rain and mud on the elfin ridgeline when she discovered one male toad, perched above a breeding pool in the thick of the high, dark forest—alone. She couldn't bring herself to touch him. Instead, she admired him from a quiet distance while the wind swept the woods around her and the golden toad sat waiting for his kin to join him on the mountain's peak.

---

* Ventana: Literally "window," the term is used to refer to the open area at the top of the Reserve where one can look down on the Pacific and Atlantic slopes and out along the ridge of the cordillera. The site is also referred to as the *mirador*: "the lookout."

Then Crump hiked on, having no reason to believe that it would be the last golden toad she would ever see.

Even as the cold and rain persisted into June and then into July, no toads emerged. Another season had come and gone, and she was growing desperate. Her study had been funded by a grant that was predicated on her observations of the mating behaviors of the golden toads, and for a whole season she hadn't even been able to locate her subjects.

"I'm thinking that the window of opportunity between dry and wet season, when the toads normally breed, simply didn't exist this year," she wrote in her field notes at the end of July. "Somehow the toads may have 'known' that conditions were never appropriate for breeding, and they're simply underground."

When Crump returned to Monteverde the next year hoping for better luck, she taped a handwritten note from her six-year-old daughter on the wall beside her bed: "I will mis you. I hop yor goldin tods come out."

At five o'clock each morning, she awoke to the alarm clock of a troop of howler monkeys in the canopy, and by seven thirty she had clambered to the elfin forest on the ridge to search for the missing toads. The depressions in the tree roots had filled with water, but the old, familiar residents were nowhere to be seen.

In the last week of April, one of Crump's graduate students from the University of Florida flew in to join the search. Frank Hensley had been in Gainesville studying for a final exam when he received the call from Crump; the wet season was descending on Monteverde, and the golden toads were sure to emerge en masse. Within a few days, he was on a plane to Costa Rica, his classes long forgotten: He was going to see the golden toads.

While Crump and Hensley waited for the toads to appear, they walked the circuit between the twelve breeding pools that Crump had noted in the 1987 season, back when they had been brimming with golden toads like overflowing cups. Now the pools were empty, so Crump and Hensley measured the water temperature, recorded the depth of the pools, checked the rain gauge that Crump had installed, and waited. They filled their data sheets with numbers, details, measurements; but their field notes only traced the outline of a ghost. The forests were empty and the toads were still missing.

In early May, Crump received a distressing call from the United States. Her husband would need emergency heart surgery, and she would have to fly home to support him and their two children. That evening, she and Hensley stayed up late into the night, reviewing the protocols for the experiments and talking through contingency plans if the weather turned: He would have to continue the search alone.

On May 15, 1989—the day before Crump's departure—Hensley climbed into the elfin forest on his own. He stopped at each of the main pools along the rugged trail, recorded data, searched the dark and dripping undergrowth, and moved on. Just past Fogden Pool—a rocky, shallow hollow named for the local photographers—he stopped in his tracks. A bright orange golden toad sat in the wet leaf litter a little off the trail, looking up at him.

"It was arguably the most thrilling wildlife experience of my life," Hensley would recall later. "It has only become more thrilling in retrospect."

Hensley had only been able to afford six rolls of film for his two-month trip to Monteverde, but he had carried his camera to the ridge that morning to photograph the breeding pools, so he laid out his raincoat on the forest floor and placed the toad on the blue backdrop to document his sighting. Not satisfied with the colorful images, he moved the toad to a mossy root protruding from a fallen log a foot away. In total, he spent no more than ten minutes with the little orange toad, assuming that it was the forerunner of what would soon be hundreds in the coming days.

"So I let the little toad go," Hensley remembers. "Wished him well. Told him I'd see him tomorrow."

The afternoon rains and swirling mist cast a veil over the mountain, and soon the ridge was hidden, and the lonely toad was left alone in the wailing, wanting dark. When Hensley returned in the following days, the toad was nowhere to be seen. It had been the last documented sighting of a golden toad on Earth.

Monteverde, Costa Rica

At the end of our first week back in Monteverde, Pri and I drove our borrowed jeep out to the Guindon farm to meet Ricky—Wolf and Lucky's fifth son. From the dirt roads that ran beyond the old cheese factory, we could see the outline of the dark mountain on the black sky, but now those forests were empty. We drove out toward the pastures above San Luis as the last light faded and the wind kicked up, sending mist in runnels through the woods where we'd heard that a puma had been hunting recently.

This road was an old, familiar path for me, which was comforting in the shadow of so much change. Since the last time I had been in Monteverde, roads that had once been dirt had been paved over, new platforms had been built over the old lookouts, more cars and buses than I remembered filled the streets. People that I'd known had come and gone. The old bars had turned into hotels. Tourism—the community's economic backbone—was continuing to change the face of Monteverde; soon, it might be someplace I wouldn't recognize.

That happens everywhere, I told myself, but that didn't make me like it any better. We reached the outskirts of the ever-changing town, and I wondered how the forests might have changed since I'd last explored them. Hurricane Nate had blown in from the Atlantic last October, pulling huge trees down and leaving open wounds in the old-growth canopy. There seemed to be more hurricanes these days, and more cabezas de agua coming down the churning rivers. Pri and I both knew people who had died with no warning underneath the landslides.

THE END OF THE SHOW

In spite of the protections that had gone up around its forests, Monteverde was a vulnerable place. It was built on the edge of existence, like a coastal town in the foam of a rising ocean. For the animals that had made their homes in its forests in the clouds, there was nowhere left to go when the world began to change around them. In the same way, the people who had called Monteverde home for generations—the grandchildren of the Costa Rican homesteaders, the descendants of the Quakers, the biologists growing old in the shadow of the mountain—were grappling with the fallout of the changing world.

I had been up late the night before, reading over Marty Crump's account of the golden toads' last vanishing act; and as we made our way toward the Guindon farm, I looked up at the mountain's misty peak and wondered how it had felt for the golden toads, to climb higher and higher in pursuit of rain and cloud, searching for the vanishing pools, only to find themselves at the top of the mountain and with nowhere left to go. How would it have felt to be a golden toad in the canopy of the world, to be the last of them, waiting in a cherished pool for a final congregation that would never come?

I knew that it was wrong to think that I could understand the feelings that had stirred within the golden toads in those dry days at the end; strange currents of time and evolution had carried us away from one another, and I could never know for sure how a toad might feel when it encountered the god of death in a dark tunnel underground. We cannot know how long the golden toads could have held out hope for the rains to come again as they waited in their hollows beneath the earth. Did they pray for rain or mourn the eggs that they had left above them in the vanishing pools? In their last days, what did the last of them remember of their golden age, drawn from buried instincts and heredity? We hadn't been there to see the toads evolve, to see them stake their claim on that high and rugged ridgeline of the continental spine; we weren't there to see them grapple with the bony hands of time—to see them climb and climb and climb until all the land was gone and there was nothing left but wind and dark and stars. But we were there to see them disappear.

I peered through the dark windshield of the jeep as we bounced along through the night, hoping that the mist might blow off and I might glimpse a vision of the ridgeline that had once been home to golden toads. I knew that they were gone, but their old breeding grounds continued to compel me; I wanted to walk the trails that Wolf Guindon and Jay Savage and Marty Crump had walked, to touch the water of the sacred pools, to see the haunted elfin forest for myself. I wanted to put it all on film, and then go home and show it to my father: *I have been where you have been*, I'd tell him. *And I have seen what you have seen.*

We didn't pass another soul on our drive; the town was battening down for the night. When we emerged over the hill, headlights cutting into fog, Ricky Guindon was waiting to meet us on the old wood porch to welcome us inside.

The house was sparse—there was a kitchen table, a couch, a mecedora* sitting before an old map tacked up on the wall, wrinkled at the corners where humidity had kissed the paper. There were family photos in little frames and epiphytes in jars of water on the windowsill, threaded together by the webs of a resident spider. My windbreaker, damp and dripping from the water in the air, left a dark pool on the rough wood floor.

"You know, the house that I was born in is just a few meters from where I'm sitting right now," Ricky told us, gesturing out into the pasture and its rolling hills. The farm was on the Pacific slope of the Continental Divide, a short walk up to the forests where Ricky had gone as a boy, with his father lighting the way, to unveil the strange bright orange toads on the high and wild ridge.

Sending his thoughts back to the enchanted woodlands and the last years of the golden toads, Ricky spoke to us softly, leaning into the arm of his creaking mecedora.

---

* Mecedora: Rocking chair.

"I started to work at the beginning of 1987 as a guide," he told us. "And in the beginning of the rainy season, when the golden toads would make their appearances, I recall seeing them a number of times. I never even thought to make a note about those observations, because they would always be there. They would be there each year."

But in the wet months of the following year, as April and May went by, Ricky would pass Marty Crump on the hidden trail up to the Continental Divide—him going up, her coming down—and he would ask her about the golden toads.

"She'd just shake her head," he said. "No golden toads appeared at all, from one year to the next."

By that time, the Quakers had been in Monteverde for more than thirty years, and they had grown accustomed to the way things changed—the old trees were swallowed by the creeping epiphytes; migrations called the bellbirds off to other forests; the golden toads descended once the rains were gone into the wet roots growing over hidden houses, darkness taking them back in—but the total absence of the familiar species was unsettling.

Pri asked Ricky if, during those early weeks of waiting, he had imagined that the golden toads might not return. He shook his head, a broad denial.

"Certainly there was no thought, at least in my mind, that the golden toads would no longer be around," he answered.

That same year that the golden toads had vanished from their diminishing pools high on the mountain, Ricky's first daughter, Hazel, was born. She would never see a golden toad—not once—in her entire life. That night as we talked together in the old Guindon homestead, Hazel moved in and out of the kitchen in her raincoat like a restless spirit, finally settling down beside her father. I had known Hazel for two years already—we had worked together; we were friends—but I had never asked her about the golden toads. Now I did; I asked her how it all had felt to her: to know of something that had been so keystone to her family's heritage and even their identity, but to never see it for herself; to hear the stories handed down from her grandfather, the way my father had passed his stories on to me. Would she tell her children all about them someday, the way that I imagined telling mine?

Hazel laughed uneasily and proposed that maybe she was actually the golden toad, reborn. She was speaking about resurrection, but my mind was on extinction. I was beginning to wonder what it felt like—when one understood that it had happened. Did that last golden toad on the high ridgeline understand its own significance? Did it mourn all that it had lost, and all that it represented in the world, as the last of them blinked out of existence like old fires dying in the night?

It may be that extinction is something that we can't really understand until it happens to us. As a species, we've spent centuries studying its work. Extinction was here before we ever were, when the world lit up and foreign stone unhusked the atmosphere, and sent the ancient titans off to roam the empty earth alone, among the mountains that burned, and one of them—just one of them, for a moment—had been the very last. It has happened in our age; we have seen it happen. It began to occur so often that we gave it a name—*extinction*, which sounds better than *annihilation*—and we began to measure it with dates, statistics, and locations so that we might put our finger on a map and say *here; it happened here, to them, not us.* We have brought it down on the heads of others, because we are a species that is able to create life, and death, and afterlife. We know about it; we might even believe that we can understand it. But we have never lived it, and as far as our evolutionary empathy might extend, we can never know what it will feel like until it happens to us. So we look to the ones who have gone before us, up ahead into the dark, and we search for signs of hope—when the moon gives no light and the last rains fail.

Ricky told us one last story that night: a story about a golden toad that he'd preserved in a Gerber jar of alcohol, in those first years when they had just begun to disappear. After a few months in the ethanol,

THE END OF THE SHOW

its golden color faded; the jar sat on a shelf collecting dust in one of the labs at the Reserve, and over time, Ricky lost track of it.

"I never found out what they did with that specimen," he said. "I always wished that they'd kept it. That would have been my own physical golden toad—to prove that it actually once existed."

On the windowsill of the Guindon house, against the glass that looked out onto a mist rising in the evening, there were seeds and stones and feathers—keepsakes carried from the forest—and a weathered photograph of Ricky's father, Wolf, taken maybe forty years ago. But there were no enduring remnants of the golden toad, no husks preserved in Gerber jars to pass down from one generation to the next.

Ricky must have seen me looking at the photograph, because he added, in a soft voice, "My father has walked other paths now. . . ."

It was a little over two years since Wolf Guindon had passed away in Monteverde at the age of eighty-five. For all the family he had left behind, there were reminders of him everywhere: the Monteverde Friends School, which he had helped to build; the Wilford Guindon trail in the Reserve, crossing over the hanging bridge; the green swells of forest sweeping up and over the divide, which he had helped protect; and the trail, long hidden, where he had once taken his son Ricky to see the golden toads—a trail that they had called *Brillante*.

"*Brillante*," Ricky told us. That was the local name for the elfin forest on the ridgeline of the Continental Divide, the name for the old abandoned breeding grounds of the golden toads—closed to visitors ever since their disappearance—and the name of the forgotten trail that led into the past. It was the trail that had been erased from the maps at the Reserve, the name we had been searching for: *Brillante*.

A map, a name, a guide. There were those who knew the way back, maybe, after all these years, and those who still remembered it; and there were some—like Ricky Guindon—who still lived in the memory of it.

Outside, the wind fell against the wooden house, but it was strong in its foundations. Ricky looked out into the vanguards of the storm and added: "It was just something that I took for granted. Something that I expected would always be there." I was no longer certain if he was talking about the golden toad or about his father.

By the time we stepped from the yellow warmth of the Guindon house and out onto the porch, the night had grown cold and damp and dark. It was on a night not unlike this one that, thirty years ago, Ricky's father, Wolf, had come pushing through a cold spring storm to knock on Marty Crump's front door with news that the golden toads had once again emerged to greet the rains. At the time, neither of them had any reason to suspect that it would be the last time they ever did.

And what was to blame for the end of the show? I began to wonder. What was it that had brought ruin to the forest of the golden toads, unlooked-for and undetected until its dark labors were already conjured? Marty Crump and Ricky Guindon had both spoken of the pools that vanished in the elfin forests; could a few dry seasons have been enough to extinguish an entire species? And what about the other disappearing frogs of Monteverde? Marty Crump had written that she'd seen fewer harlequin frogs and glass frogs in 1988 as well. I didn't know the identity of that creeping fear, but I began to wonder if the golden toad had been its only victim.

And something else that Ricky said was gnawing at me. Pri had asked him how it felt to find the corpses of the golden toads, and Ricky answered that he had never found a body—no one had. Later that night, I would go back to the field notes Marty Crump had written during her last season in Monteverde, and I would realize he was right. There were no reports at all of dead or dying golden toads, no bodies scattering the bright pools of the elfin forest. They had simply vanished.

I pulled my hood up to keep the cold from cutting into me while Pri talked to Hazel and finished her cup of tea in the doorway of the old house, perched on a thin ledge between the past and an uncertain future. Above us, the canopy swayed back and forth in the eternal wind, silhouettes against the deep black sky. I took out my phone and typed two sentences to Kyle; it was late, and he would be asleep, but I wanted him to see the message first thing in the morning.

We need to find out for sure what happened to the golden toads, I wrote to him. *I might need you to help me catch a killer.*

As we walked to the jeep and climbed inside, the dying headlights lit the Guindon house against the backdrop of the purple night, where we could see Ricky rearranging furniture through the window, setting his house in order. I backed the jeep up and we pulled away into the dark. As we began again down the muddy roads, the lonely moon spilled its white glow onto our path, which was clearer now than it had been at any point on our long journey. We had a map, and Ricky had given us a name: *Brillante*, the old trail into the haunted breeding grounds of the golden toads. Pri and I looked up at the dark ridge on the drive home. It would be a long and restless night.

Soon we would walk into the bastion of the golden toad.

CHAPTER FOUR
THE CREEPING FEAR

Canterbury, England

The hallways of the University of Kent in Canterbury were alive with rumors. In September 1989, the First World Congress of Herpetology had drawn close to fourteen hundred herpetologists from sixty-one countries to the university, and over the nine days of workshops, round tables, and symposia, they milled about in small knots to swap accounts from their field sites; ripples of anticipation ran through the crowds whenever a young researcher shared a preview of unpublished data, or a new theory reached the ears of the old masters. Colleagues who hadn't met in decades reunited to unveil the scientific revelations they'd brought out of the jungle, comparing notes and scars and strange encounters. It is unlikely that there were more people anywhere else on the planet talking about frogs.

But amid the excitement and enthusiasm, a dark cloud was gathering over Canterbury. Word had spread of a herpetologist who had brought a tale of grief out of the tropics—a mystery of lost gold and empty forest, a vanishing act that would stump the best magician.

Marty Crump was not scheduled to formally present on her research at the Congress, but her worry for the missing golden toads had been gnawing at her ever since she'd left the country four months earlier. Back in May, her graduate student Frank Hensley had reported his sighting of a single toad in the Brillante breeding grounds, but all the other searches had proved fruitless. Crump had not seen a golden toad herself since May 1988, and it had been two years without a normal breeding season. It was clear that something was amiss.

THE GOLDEN TOAD

On her first day in Canterbury, she told a few friends about her strange experience in the elfin forests and the disappearance of the golden toads, and it didn't take long for word of the lost species to spread. Some of the herpetologists who sought Crump out only wanted to hear the unsettling story, but others were looking to make confessions of their own: Piece by piece, a chilling pattern began to emerge. Cynthia Carey, a physiological ecologist at the University of Colorado in Boulder, had found the carcasses of once-plentiful boreal toads (*Bufo boreas*) outside the ghost town of Gothic as early as 1973; by 1976, they seemed to have vanished from her old study site entirely. Stanley Rand, from the Smithsonian Tropical Research Institute, was in the midst of investigating a dramatic amphibian population crash in Boracéia, Brazil; since 1979, six of the thirty common frog species had disappeared, and when in 1982 Rand had returned to his old study site for the first time in almost twenty years, he was haunted by the distant calls of endemic tree frogs that he could hear but never see.

Michael Tyler, who had come to Canterbury all the way from the University of Adelaide, had carried a few ghost stories of his own. Australia was in the midst of an enigmatic amphibian decline unlike anything its scientists had ever seen. In 1973, Tyler had been among the biologists who first witnessed the novel reproductive process of the newly discovered southern gastric-brooding frog (*Rheobatrachus silus*), which apparently swallowed her own fertilized eggs, brooded the young in her stomach, and then gave birth to them through her mouth. It was one of the most important herpetological discoveries in recent memory—the practice of gastric brooding was so unbelievable that the journal *Nature* had rejected the groundbreaking paper as a hoax—but less than ten years after its initial discovery, the new frog had seemed to vanish. In 1979, one of Tyler's contemporaries was forced to place his study of the gastric-brooding frog on hold when the species disappeared from the area without a trace. At the same time, other Australian frogs of previous abundance were also declining. The

THE CREEPING FEAR

southern day frog (*Taudactylus diurnus*) was normally so numerous that anyone camped beside a stream was likely to be woken up by dozens of day frogs hopping all around them—but 1979 had marked the last sighting of the day frog too. Australian biologists were at a loss; some noted a general decline of animal life due to a consistent lack of spring rains, while others saw nothing to be concerned about in the apparent "disappearances."

But someone who was concerned, and who was also in attendance at the World Congress in Canterbury in 1989, was David Wake, the director of the Museum of Vertebrate Zoology at the University of California in Berkeley. He had recently traveled back to the pine forests near Oaxaca, Mexico, where a decade earlier he had found more than eighty inch-long salamanders beneath the bark of a single log. On the return trip, he had searched all day to find just two specimens.

Of all the strange reports making the rounds at the World Congress of Herpetology, Wake's experiences might have been the most concerning. He had also witnessed the disappearance of the mountain yellow-legged frogs (*Rana muscosa*) in California's Sierra Nevada; hiking near Tioga Pass in the summer of 1959, Wake had seen so many of the spectacled amphibians that he'd had to watch his boots to keep from stepping on them. The biologists Marc Hayes and Mark Jennings had sounded early warning bells in 1986, and three years later, Wake heard from David Bradford at the University of California in Los Angeles that he had managed to find yellow-legged frogs in only 2 percent of the lakes he'd monitored for the last ten years. Just before traveling to Canterbury for the Congress, Wake had returned to the Sierra Nevada to confirm the tidings for himself: All of the adults had vanished—only a handful of tadpoles remained.

When Wake heard rumors that the famous golden toads of Monteverde had disappeared as well, he sought out Marty Crump in the crowded corridors of the University of Kent between conference sessions and presentations. He told her about the salamanders in Oaxaca and the yellow-legged frogs in the Sierra Nevada, and one more story that hit close to home for both of them: In 1987, Wake had visited Monteverde to study a local species of *Bolitoglossa* known as the Monteverde salamander and found dozens of them in the cloud forests above the small community. When he'd returned just one year later, he couldn't find a single one. "Hard to understand," he'd written in his field notes in 1988; but in Canterbury, England, in September 1989, the vanishing act felt less bewildering than it did portentous.

With these reports in mind, and with a cold chill running down her back, it occurred to Marty Crump that the golden toads were not the only frogs she had witnessed disappear from Monteverde. Between November 1982 and April 1983, Crump and her graduate student Alan Pounds had counted more than seven hundred black and yellow harlequin frogs on the banks of the Río Lagarto four kilometers southwest of Monteverde; even in the drier weather, they had found as many as two hundred frogs some days, sitting out in the open on boulders and logs or partially hidden in crevices between the rocks. But when Crump returned to the same site in July 1988—shortly after encountering the last golden toad she would ever see—she searched for six hours before finally giving up. There were no signs of Monteverde's harlequin frogs. On the evening of July 6, Crump had taken out her field notebook to record her apprehensions, for the first time giving names to her fears. "What's happened?" she wrote. "Have I done something?"

She hoped that she was wrong, that she had not played a role in the disappearance of the very creatures she had come to study. Maybe it had been something else entirely—something completely out of her control. She considered the El Niño drought from the year before, when leaves crunched underfoot and the ephemeral pools dried early, leaving behind desiccated eggs and week-old tadpoles with no hope of survival. The Arenal volcano had also been erupting violently for the last few years, spewing ash and noxious chemicals eighteen kilometers to the north; could pollution from the volcano be to blame? She recalled the rumor that the golden toads were advertised in black-market European pet trade magazines for as much as $500 apiece; could collectors have descended on the elfin forests? "But when could they have done this?" she wondered. "Somebody would have seen them. And anyway, there's no way every last golden toad could have been snatched up and smuggled out."

Now, cast in the context of the other accounts she had heard at the First World Congress of Herpetology, the disappearance perched atop the others like a house of cards. To Marty Crump, and to many others in attendance, it was becoming clear that the case of the missing golden toads was not an isolated incident—no crime of passion, no thief in the night. Something was at work on a global scale; and if the available evidence was to be believed, it was already well on the way to an amphibian annihilation, and it was possible that they were already too late.

Back in July 1988, as she had sat in the shadow of uncertainty with no harlequin frogs to console her, Crump had written a brief musing in her field notebook as she considered the cause of the frogs' absence.

"Maybe," she had written, "a local outbreak of a particularly nasty parasite that attacks and kills frogs?"

Washington, D.C.

In September 1996, a veterinary pathologist named Don Nichols received the news that he'd been waiting to hear for the last five years: The frogs were dying again.

His vigil had started in 1991 when, unsatisfied by the stimulation of his day job conducting postmortem exams on mice and rats at the National Institutes of Health, Nichols had been moonlighting as a veterinary pathology consultant, mostly helping friends and neighbors investigate the afflictions of their pets. One day, a colleague at the University of California in Santa Barbara mailed Nichols a package. He opened the box to find three dead arroyo toads (*Bufo microscaphus californicus*) and a short but alarming note: In just two months, 60 percent of the university's captive population of the endangered toads had died from an unknown cause.

Intrigued by the mystery, Nichols did what he could to diagnose the deceased toads: He studied samples under microscopes, noticing "numerous microscopic single-celled organisms unlike anything I had ever seen before." He sent skin sections off to experts in animal disease, but every partial answer seemed to raise more questions. Whatever was at the root of the deaths of the toads, it was something that the scientific community had never seen before. And with the source colony dead and gone, there was no way to get his hands on any other samples. So Nichols put the box away—out of sight, but not forgotten. When he was offered his dream job at the National Zoo in Washington, D.C., that fall, one of his first orders of business was to scour the Smithsonian files for anything resembling the skin infection that he'd seen on the arroyo toads. His search paid off: He discovered an identical case in three frogs from the Zoo's collection. But there was little more information to be found, and no indication of what examinations, if any, had been conducted on the specimens. It was another dead end. Frustrated and mystified, Nichols resigned himself to patience; whatever this unknown killer was, it had been here before: Maybe it would strike again. For the next several years, Nichols had a standing order for his pathology residents: Whenever an amphibian was presented for necropsy, they were to collect skin samples for examination. Whenever he was asked why, his reply was the same: "Because I am looking for something and I don't know what it is."

He wasn't alone: Ever since the First World Congress of Herpetology had sparked an international awareness of the world's disappearing frogs, scientists across the globe had been turning over the mystery, chasing down leads, and naming suspects. A slew of articles began to fill scientific journals and popular magazines alike, with titles like "Where Have All the Froggies Gone?," "The Case of the Disappearing Frogs," and "Why Are Frogs Croaking?" Study after study propounded different hypotheses—acid rain, pesticide poisoning, harmful UV radiation from ozone depletion—but none of these could account for the global synchronicity of the declines.

Then, in September 1996, Don Nichols's intern Allan Pessier came bearing the news that Nichols had been waiting for: The Zoo's blue poison dart frogs were dying.

Pessier, a young veterinarian, had only recently begun his residency at the National Zoo, working under Nichols to monitor the captive amphibians for death or disease, and performing autopsies on every frog and toad that died at the Zoo. It was during the course of these usual proceedings when a wholly unusual number of blue poison dart frogs (*Dendrobates azureus*) began to cross his desk.

"It had been a successful program for years," Pessier said. "But all of a sudden, everything was dying."

Bringing his findings to his mentor, Pessier watched as understanding dawned on Nichols's face.

"Don said, 'Oh my God—I've seen this before.'"

In the terrariums and displays beyond the lab, dozens of frogs sat slumped lethargically at the edges of their tanks, past the point of no return. Their skin had taken on a brownish-red discolored hue; in places it was already beginning to peel away. The frogs' dark eyes peered back through the glass in silent pain, seeking mercy.

With fresh samples to work with, Nichols could finally pick up where he'd left off; as the dying frogs looked on, he searched his bookshelves and pulled down a well-thumbed textbook,* opening it to a bookmarked page. Earlier that very year, he'd been working on a presentation for the annual conference of the American Association of Zoo Vets. Its topic: the mysterious skin infection from the box of arroyo toads.

But as they paged through the textbook together, Nichols and Pessier found nothing that matched the images they'd taken in electron

---

* The textbook was *Handbook of Protoctista*, Margulis et al., 1990.

micrographs of the dying dart frogs. The closest they could get was a category of zoosporic* fungi from the ancient phylum *Chytridiomycota*—but there were no records at all of a chytrid causing a disease in vertebrates. Nonetheless, nothing else looked close; it was the only lead they had.

Both Nichols and Pessier were well versed in the usual suspects of veterinary pathology, but chytrids were far beyond their areas of expertise. After several hours on the Zoo's text-only web browser, Pessier stumbled upon a webpage entitled "Zoosporic Fungi Online," which listed the contact information he was looking for. He carefully prepared and scanned one of the electron-micrograph images of the skin infection, along with a hurried email seeking help. *What ails them?*

Soon after, Pessier sent the electronic distress call off like a message in a bottle to the laboratory of a mycologist named Joyce Longcore, and retreated to the halls of frogs and toads to wait.

Today, Longcore still keeps those original electron-microscopy photos in her office at the University of Maine. On a cold, wet February day, she drew them out from a box of overflowing papers to show Kyle when he met her within Deering Hall. That morning, Kyle had left his wife and cats sleeping in their house at the edge of the woods on Boston's north shore to drive I-95 north along the Atlantic coast to Orono, where the University of Maine's campus huddled among the bare trees, trying to keep warm amid an endless rain.

For some time now, Kyle had been on the trail of the golden toad's killer, immersed in the dusty shelves of libraries, tracking down out-of-print articles, mining the long-sheltered memories of those who had been there to see the beginnings of the amphibian crisis unfold. At first, Kyle had managed to keep a safe distance from the mystery that had taken ahold of me; but as time ran on, its shadow had begun to fall over him as well. When we'd first met up again after he'd gone west and I'd

---

* Chytrids distinguish themselves from other fungi by their flagellated zoospores, whip-like tails that allow these agents of asexual reproduction to propel themselves through water or even moist soil.

gone south, I had told him the story that I'd heard—the golden toad, its disappearance, the unanswered questions that had piled up. The golden toad had not seemed that important to him; he had always been more of a birder. But bit by bit, he'd begun to send me journal articles he'd found, emails asking questions, text messages about other frogs that had disappeared. By the time he'd finished grad school, he, too, had been called onto the trail of the golden toad—or, more specifically, its killer. And in Joyce Longcore he had found a key piece of the puzzle.

Deep within the catacombs of Deering Hall, he sat down with the eighty-five-year-old mycologist who, for the last forty years, had been one of the world's leading experts on chytrids—an obscure branch of mycology that until the late 1990s had been afforded little interest and even less funding.

There, within the narrow office, the gray-haired, smiling professor took out boxes overflowing with folders and papers; and then she passed across the desk the glistening photo paper—the original electron-microscopy photos that Allan Pessier had sent her.

"It was diagnostic," Joyce Longcore told Kyle as they sat together over a homemade lunch she had prepared for the two of them amid the stacks of papers and books and boxes. "It was a chytrid."

As soon as she had realized what she was looking at, she had also known what it could mean. Through images alone, however, Longcore knew that she couldn't draw a firm conclusion. She had immediately replied to request fresh tissue, sending Pessier off to prepare new samples from the blue poison dart frogs—he packed a disembodied leg in ice, and sent it north to the University of Maine. Longcore then set about trying to isolate and grow the specific chytrid in the controlled environment of her lab. The results were disheartening; each growth medium she tried proved unsuccessful. It wasn't until October 13, 1997—her birthday—that she achieved a breakthrough. In her last attempt, one tiny piece of frog skin on nutrient agar had been seen to host several chytrid bodies—but they hadn't developed further, and so Longcore had

THE GOLDEN TOAD

finally placed the initial growth into a flask of liquid medium and put it aside in frustration, leaving it alone on the lab bench. She had nearly forgotten about them when she came into the office and caught a glimpse of an opalescent[*] liquid out of the corner of her eye: Against all odds, her improvised growth concoction had worked. When she examined the new chytrid more closely, Longcore realized that not only had she inadvertently invented a successful procedure for growing the culture, but she had discovered a new genus entirely. With no one else in the lab with whom to share the news, Longcore recalls, she exorcised her nervous energy by zipping around the empty hallways in silent celebration.

Back at the Smithsonian National Zoo, Don Nichols and Allan Pessier received Joyce Longcore's news with a mixture of excitement and trepidation: For the first time, the culprit behind the frog mortalities was coming into focus. With Longcore's isolate, they followed Koch's postulates—a series of criteria designed to confirm the link between a microbe and disease.[†] And as more of the Zoo's frogs perished, one by one, the reality became impossible to ignore: This skin fungus—this chytrid—was the killer they'd been searching for.

But they were not the only ones on the trail of dying frogs. In September 1997, one of the reptile keepers at the National Zoo walked into the lab and dropped a copy of the *New York Times* on the desk in front of Nichols and Pessier. The headline read: "New Culprit in Deaths of Frogs."

---

[*] When Kyle asked her about the nature of this opalescence, Longcore shook her head, at a loss to explain it. "I've never seen it again."

[†] The criteria are as follows: (1) The bacteria must be present in abundance in every case of the disease and must not be present in a healthy organism; (2) the bacteria need to be extracted from the host and grown in pure culture and identified; (3) the bacteria are inoculated back into a healthy host and must then cause the onset of the disease; (4) the bacteria must be extracted from the inoculated host and grown in pure culture to be identified as the original causative agent.

THE CREEPING FEAR

Fortuna Forest Reserve, Panama

The sliver of a crescent moon—offering little light and less hope—had gone down into the soft bed of canopy, and the jungle had settled into darkness. The biologist Karen Lips moved alone through the waiting hour just before first light, wraithlike, among the thick understory of silent Fortuna. As she walked along the banks of the Quebrada Chorro, the long-poled net she carried dripped with cool water from the little stream where she had been netting tadpoles. She was deep in the Reserva Forestal de Fortuna on Panama's Caribbean slope, which spanned a forest that climbed from sea level up to fourteen hundred meters; a pristine, unspoiled habitat like a green jewel atop the watershed of a sweeping river valley. In the dark, the jungle would come alive with the sound of a thousand voices singing heart-songs from high limbs, or trilling secrets from the wandering streams. But on this January morning in 1997, Lips walked from stone to stone beneath a cloud of fear; the hoped-for calls of the *Colostethus* frogs that had once been common among the hanging leaves above the stream were nowhere to be heard. For the past few weeks, the forest had been growing quieter, and now it was nearly silent—as if it had been abandoned. Trudging along the ragged trail, warding off the low branches and creeping roots, she squinted up the path in the scattered light of her headlamp, stood listening, heard nothing. Then she left the banks of the Quebrada Chorro and struck off toward the Quebrada Arena. In the gulf of quiet jungle, uncertainty began to gnaw at her, a tigre chewing bones: Would she find more of the same up ahead? More absence, more nothing? These were ethereal times, when the discovery of nothing had become important. But when she emerged onto the banks of the Quebrada Arena, stepping through a screen of hanging cecropia to look upon the singing stream, she encountered a different scene entirely: the scene she had been chasing for four and a half years.

Back in 1992, Karen Lips had just started her Ph.D. work in the cloud forests of Las Tablas, Costa Rica, not far from the Panama border. Her adviser at the University of Miami—none other than Jay Savage—had sent her off into the mountains of the rainforest where she had been living for the past several months in a four-by-four-meter shack bereft of running water and electricity, which she had christened "La Casita." Each morning, Lips would rise to hike two hours through the rugged trails to the top of the watershed to study the breeding behavior of spiny tree frogs (*Isthmohyla calypsa*). Metallic emerald, the little frogs looked like green and gray lichen on a mossy branch. Beneath the silent gaze of resplendent quetzals—and other more carefully concealed observers—Las Tablas was like a wild holy garden untouched by calamity.

"I was living a field biologist's dream," she would write later. "I couldn't know that my study site in this remote cloud forest would give me a front row seat to one of the most distressing ecological mysteries of our time."

In December 1992, Lips left the tropics to go home to the United States for Christmas. When she climbed back up to Las Tablas at the beginning of the rainy season in the new year, it was as if she had entered a different world: The frogs simply weren't there. At first, she soothed her concerns with rationalizations about the weather; it was drier than usual, and the rains hadn't returned yet in their full abundance. As she settled in to wait, doubt crept in: Could her sampling technique have caused the disappearance? Had the bright light of her headlamp scattered the frogs? Here and there, she would find a wandering frog or two along the streambanks, but they were dead or dying: They would jump once, and then expire on the ground in front of her. It was as if something had clambered over the walls of the garden in the night while she had been away, undetected and unlooked-for: a silent killer on a mad crusade.

She made a list of theories and one by one disproved them to herself. Habitat destruction, introduced fishes, UV-B radiation, acidification: None of these made sense. The killer was behaving like a chemical or a pathogen, maybe emboldened by environmental contamination or climate change. The days went by and a darker thought began to haunt her, like a hunting cat creeping up her trail. She had heard about the golden toads of Monteverde—just a few hundred miles northwest of where she stood—and how they had disappeared, from one season to the next. In the long, dark nights on the mountain in Las Tablas, with La Casita moaning in the wind, she began to wonder if she was watching something migrate through the forests of Central America. Whatever this phantom killer was, it seemed to be on the move.

But if there was an answer to be found, it wasn't going to be uncovered at Las Tablas. Here, the killer had already been at work; the frogs were gone, vanished or devoured, dead or dying. So Karen Lips decided to play a hunch: She moved southeast, across the border and into Panama, and set up camp at the Fortuna Forest Reserve. There, on the backbone of the cordillera, she hoped that she would find abundance: a cloud forest untouched by the calamity that had brought such grief to Monteverde and Las Tablas.

At first, the forests of Fortuna didn't disappoint; when she first settled into her fieldwork in Panama, she was almost tempted to forget about the silent jungle she had left behind in Costa Rica. The Fortuna field station had electricity and warm showers, a paved highway crossing dozens of streams, and historical survey data of forty frog species that all still seemed to be abundant in the river valley. She set up new transects and started a new study, grateful to have another forest full of frogs; but below the surface, she was still waiting. And it was there at Fortuna, in January 1997, that Karen Lips met a killer.

Under the faint light of the crescent moon, the Quebrada Arena had become an open-air graveyard of frogs. As she pushed aside the sheltering cecropia and made her way to the stones beside the cool water running in the early morning light, Lips knelt down to examine the little corpses littering the streambed and the understory. One by one, she found them: dead frogs arrayed like sky-fall, silent witnesses to the end of their existence. Many of the frogs were frozen in their normal calling postures, as if they had come down to the streamside the night before and been caught unawares; you might imagine they had died calling for help.

Lips took a step back—she didn't want to contaminate the scene—and crouched down, breathing fast and squinting through the morning's hazy light. She began to count the dead frogs on the streambank from a distance—eight, ten, more than a dozen—and then a twitch of movement caught her eye among the rocks in front of her. She moved closer. There, on a wet rock at the edge of the water, was a lone survivor. The frog made no effort to avoid her grasp when she reached for it, soft hands cupping it in darkness, and Lips could feel the small frog's heart beating fast but tired through its moist skin against her own. Drawing it close, she lifted up her covering hand and looked into the eyes of the little frog.

It is possible that the small frog looked at the world through clear eyes in those moments before its death; that it was granted a kind of foresight or clarity that is otherwise withheld from us. It is possible that it understood in one way or another that this colossal interloper, like a titan from another world, had come to this forest to try to save it; to understand what had happened to the others, if she could. The small frog waited bravely in her palm as she looked him over, his body glistening as if anointed, though he trembled with a whisper of fear, and an incomprehensible intuition that the world was coming to an end.

Lips observed no wounds or lesions on the skin, but she knew that there was more to be discovered under bright light and careful gaze back at her lab. She would take the small frog with her to the field station, along with the others whose lifeless bodies hid the answers to the questions she'd been chasing.

She would need her hands free to pack him safely, so she lowered her hands to the rock to allow the little frog to crawl free. But he merely sat there in her open palm, not moving; his trembling had stilled; his tired heart had stopped.

THE CREEPING FEAR

As Karen Lips wandered back through the silent forest toward the field station at Fortuna, she paused long enough to look back the way she'd come at the Quebrada Arena, the thin places where the water moved between enclosing trees. It was not only a graveyard—a hallowed place where the memories and ghosts were young—but a crime scene too. For three and a half years, since Las Tablas, she had been on its trail; here at Fortuna, it had struck again. Monteverde, Las Tablas, Fortuna: three remote, protected, high-elevation forest reserves; three sites of sorrow and annihilation. As she climbed the steep trails through the misty forest, Lips knew that she had something now that others hadn't: evidence.

Quickly, Lips sent a handful of the corpses to a veterinary pathologist in Maryland named David Green, who identified a skin infection as a likely factor in the mortalities.

"I could find no evidence of widespread viral, bacterial, or fungal infections," Green reported to the *New York Times*. "But it looks like a protozoan of some sort had infiltrated their skin."

Reading the story in the office at the National Zoo, the air heavy with the smell of formaldehyde, Don Nichols and Allan Pessier could feel the eyes of their lost poison dart frogs on them; they were floating in jars on the shelves above, not so different now from ghosts. It was because of those frogs that Nichols and Pessier knew something at that moment that the rest of the world did not: This was no protozoan.*

The *New York Times* article also referenced an unnamed Australian pathologist, who had reported to David Green that she was seeing the same skin infection in dead frogs fifteen thousand kilometers away.

---

* To Don Nichols, this mistaken identity was clear even within the evidence the article presented. "This article even has a photomicrograph of frog skin in which the 'protozoa' look like chytrids," he wrote to Joyce Longcore. "You can even see the rhizoids!" However, due to the slow-paced nature of print publication, this statement of singular knowledge is not entirely true. While the *New York Times* story went to print on September 16, Lee Berger (the Australian pathologist whom we are soon to meet) had one week earlier received an email from Louise Goggin of the CSIRO ruling out the suspected identity of a protozoan and suggesting instead its close resemblance to a flagellated fungus: a chytrid.

Now her Australian team was planning to convene with David Green and Karen Lips in the U.S., where they planned to review their newest Australian and Panamanian samples alongside older preserved specimens to combine their disparate pieces of a global puzzle: the dispersed clues that all pointed to the same killer.

This might be more important than we thought, Pessier remembers thinking.

What he said was: "I think we need to get invited to this meeting."

Urbana, Illinois

A coterie of unlikely collaborators gathered at the University of Illinois in Urbana on a cold October day in 1997 to look a killer in the eyes. From Central America had come Karen Lips—the young researcher who had reported a sudden population crash of Panamanian frogs. David Green came with her—the pathologist who had examined the samples she had carried home from Panama. From the United States came Don Nichols, Allan Pessier, and Joyce Longcore—the veterinary pathologists from D.C. and the mycologist from Maine who had identified the killer as a chytrid fungus, and successfully grown it in the lab. And from the Australian team, who had been among the first to sound the alarm of a pathogen, and to implicate it red-handed in the annihilation of their captive frogs, came Lee Berger.*

Since September 1995, Berger had been working as a Ph.D. student at James Cook University under the guidance of Keith McDonald and Rick Speare, who were investigating an unidentified frog disease ravaging local populations. McDonald, of the Queensland Department of the Environment, had his own ghost stories to tell. He had been the last to see the southern day frog in the wild, in 1979, and since then he had closely followed the news articles published by two other Australian

---

* With her were her associates Andrew Cunningham and Peter Daszak, both of whom were instrumental to the Australian efforts.

biologists*; their stories, like "The Mystery of the Disappearing Frog" and "The Twilight Zone," alluded to an unidentified "Event" that had orchestrated the overnight disappearances of the southern day frog and the southern gastric-brooding frog. For three years, McDonald had been combing Australia's D'Aguilar, Blackall, and Conondale ranges in the hope of finding some trace of whatever had spirited away these frogs. It was during the course of these excursions that McDonald partnered up with Rick Speare.

A veterinarian and medical doctor, Speare was a promising accomplice: He had previously studied disease agents in cane toads, and together the two men began to suspect an infectious disease spreading north.† To test their theory, they established a study site at O'Keefe Creek in the wet tropics of Big Tableland, the last known habitat of a small amphibian called the sharp-snouted day frog (*Taudactylus acutirostris*). Within a few months, their suspicions were proven correct: They arrived to find little more than scattered corpses of sickened frogs.

In 1994, Speare collaborated in the founding of Australia's Amphibian Disease Team and led their efforts to examine fresh specimens, ruling out previous hypotheses like bacterial septicemia, aeromonads, and ranavirus—but they were unable to isolate a specific causal agent, a single culprit. All that they could note with certainty was a series of abnormal skin lesions along the frogs' backs, which were long-persisting in the captive populations despite intensive care. Although he couldn't know it yet, Speare was looking their killer in the face. What Speare did know was that these lesions, combined with the geographical pattern of the disappearances, confirmed their theory of an infectious disease; but a paper published with McDonald in the journal *Conservation*

---

* Glen Ingram and Greg Czechura were two early and passionate voices calling attention to the Australian amphibian declines, and lobbying for conservation of the Australian rainforest. It was Ingram who had to put his research study on hiatus when the gastric-brooding frogs disappeared, and Czechura was one of the last to see these frogs in the wild.

† They were arguably the first to do so. As early as 1989, Rick Speare had isolated a virus from an ornate burrowing frog (*Limnodynastes ornatus*) in northern Australia.

Biology would be met with skepticism and detraction from the scientific community.

Confident that they were on the right path, McDonald and Speare determined to call in the support of a full-time veterinary pathologist; because of the academic resistance to their theory, they were only awarded enough funding for a single Ph.D. student—the young veterinarian Lee Berger.

At that point, Speare suspected that the infectious disease they were dealing with had to be caused by a parasite or a virus; when Berger joined the team, she went to work studying the declines that were now widespread across Australia; but she struggled to draw conclusions from the available specimens.* What they needed were fresh corpses, but they had no idea where to find them; McDonald and Speare had been lucky once at Big Tableland, but lightning wasn't likely to strike twice. By the end of 1996, Lee Berger had begun to wonder if they would ever catch the killer in the act.

In the end, it was luck and good timing that brought Berger the fresh samples she needed.

On a sunny day at the end of 1996, Lee Berger lit out for Melbourne to meet a local breeder of endangered frogs named Gerry Marantelli, whose frogs were dying by the hundreds. When she reached his property, Marantelli beckoned her inside, and they stood together peering in through terrariums and tanks at dozens of dying frogs: lethargic and sick, draped in the same discolored and sloughing skin that Berger had seen time and time again inside the lab. But those samples had been weeks, months, sometimes years old: These were fresh, wearing all their secrets. In the shadow of the death that was soon to overtake these frogs, Berger felt her hopes, once so heavy, finally beginning to rise. Here at

---

* The skin samples that had been examined from the die-offs at Big Tableland were mostly pieces from the backs of the dead frogs—as it turned out, probably the worst place from which to sample. "I think if Rick had taken the ventral skin," Berger considers in retrospect—the skin from the underside of the frog—"he'd have worked it out straight away."

last was a teeming supply of fresh infected tissue: Here at last were the clues she sorely needed.

With Marantelli's blessing, Berger collected samples of infected tissue from his dying frogs and took the precious cargo back to her lab in Geelong. There, at the end of February 1997, she infected six great barred frogs (*Mixophyes fasciolatus*) by adding skin scrapings from the sick specimens to the water of the healthy frogs. One frog died; the rest were euthanized when it grew clear that they would soon succumb to the same disease. From those sacrificed frogs, Berger drew the knowledge that they had all been searching for: They had finally found their killer.

But by the time the emissaries gathered from across the globe in the windswept campus at Urbana, Berger had grown aware of a chilling truth that none of them had yet the heart to speak aloud: In the time that it had taken them to realize what was happening, this chytrid fungus had become globally dispersed. Now they had two tasks in front of them: to share their stories with each other, offering what they knew; and to confirm beyond the shadow of a doubt the identity of this phantom menace.

At first it seemed like tension and uncertainty might threaten the fellowship that the moment called for, and for Lee Berger it was a lesson in the challenges of scientific collaboration. After so many years of laboring in solitude—struggling against dismissal, detraction, and doubt—the independent contingents were in some degree reluctant to reveal all the fruits of their thankless work, worried of being scooped in their discoveries, their relentless efforts downplayed or overlooked; sentiments of "have fun in Urbana and don't give anything away" are known to have been spoken in more than one instance.

It was George Rabb (then the director of the Brookfield Zoo and the man who was largely responsible for securing funding and organizing the gathering) who addressed the reticence head-on. "He got angry," Pessier recalls. "He got up and said: This isn't about *Science* or *Nature* or anything else—this is about the frogs, damn it."

This condemnation seemed to break the tension, and those in attendance began to see each other more as allies and less as rivals; after all, they had all endured the skepticism from some parts of the scientific community—had their theories shouted down or else ignored entirely. And here at last they had the evidence they needed to shine a light upon the thing that had so long endured in darkness.

Beneath the university's powerful electron microscope, the teams examined slides of infected frog tissues from disparate sites and specimens. With their collected knowledge, they agreed that this same causal agent—this chytrid—was indeed present in the lost frogs of Australia and the Americas, and that it was the proximate cause of the disappearing frogs that they had each observed in their own countries. The killer had come out from the shadows, in the end—it stood before them, terrible to behold.

When the assembly dispersed, their work was far from over, and their collaboration would continue. They planned to publish their corresponding research papers side by side in the same scientific journal to emphasize the collective efforts that had made their revelations possible. *Nature* rejected both papers.* So, in July 1998, Lee Berger published her dissertation in the *Proceedings of the National Academy of Sciences*; among the coauthors were Rick Speare, Keith McDonald, Gerry Marantelli, David Green, and Karen Lips. The paper identified chytridiomycosis as the cause of amphibian mortality associated with the population declines in the rainforests of Australia and Central America. And in April 1999, Joyce Longcore, Don Nichols, and Allan Pessier published their findings in *Mycologia*, the journal of the Mycological Society of America. In honor of the blue poison dart frogs from the genus *Dendrobates* who had given their lives to reveal the culprit, the group named the fungus *Batrachochytrium dendrobatidis*.

---

* Effectively quashing the intention to publish their papers side by side, this rejection left the two groups to seek publication wherever they might find it. As a result, rumors of contention would long hang unnecessarily above the heads of these contemporaneous discoveries.

THE CREEPING FEAR

And like that, the creeping fear had a name.

Atlantic Coast, Maine

By the time that Kyle left Joyce Longcore's lab at the University of Maine, walking out of the stuffy halls into blowing snow, he'd been long on the trail of *Batrachochytrium dendrobatidis*. He'd heard from the biologists in search of lost frogs in Australia, from the veterinary pathologists and mycologists examining the zoosporic fungi in the United States, and from the scientists and students who had spread out their puzzle pieces together on the floor, finally giving a name to the long-sought killer responsible for the decline of frogs across the globe.

But was the golden toad among them?

That was the question that Kyle began to ask himself as he drove south along the gray Atlantic coastline. As the storm began to break apart, he made a detour to the beach, walking out along the boundary between earth and sea like a spaceman at the edge of gravity's reach. It was the same beach that he and his wife often went to in the summer; but now it was colorless and cold, and it had the feel of winter, the end of things, all about it.

The chytrid fungus was a proven frog-killer, and its methods fit the profile of the golden toad's disappearance; but the evidence in this specific case was circumstantial. Had *Batrachochytrium dendrobatidis* had a hand in the disappearance of the golden toad? Or was there still another killer to discover, long at work in the green hills of Monteverde, that had lent its dark designs to the golden toad's annihilation?

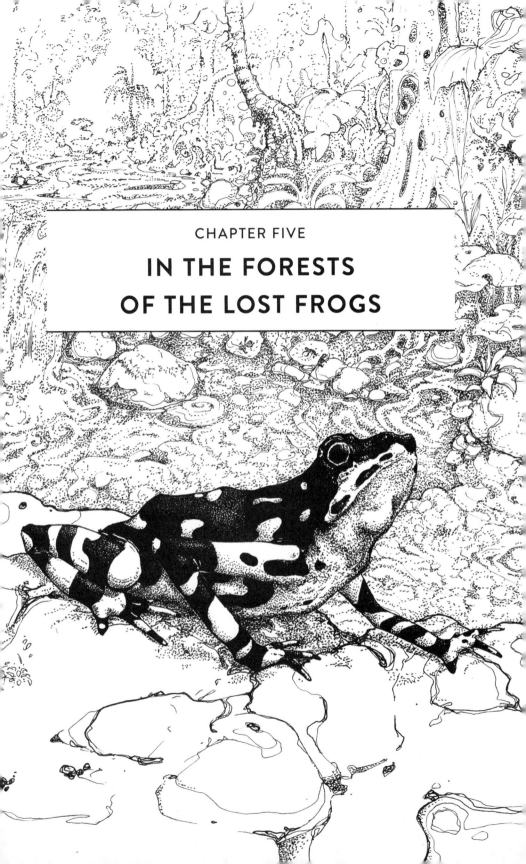

CHAPTER FIVE
IN THE FORESTS OF THE LOST FROGS

Monteverde, Costa Rica

Three hundred Fleischmann's glass frogs (*Hyalinobatrachium fleischmanni*) had once held territories along a 120-meter stretch of the Quebrada Cuecha in Monteverde, but the banks were quiet now as Alan Pounds crept along the little stream, making note of yet another victim.

By 1990, news had reached Pounds from his former professor Marty Crump that the golden toads were missing from the highlands and the harlequin frogs had vanished from their previous study site. So, working closely with the local photographer and naturalist Michael Fogden, Pounds had set off to conduct a series of ambitious surveys across the Continental Divide. He'd searched through boundary pastures and into primary forest, along streams and through swamps, up into hidden forest trails. The observations he recorded were alarming: He found no variable harlequin frogs, no red-eyed leaf frogs, no brilliant forest frogs, no bare-hearted glass frogs, no green-eyed frogs, no golden toads. Out of the fifty species of frogs and toads that had once been common in Monteverde, twenty-five were missing.

The same year that he began the local amphibian surveys, Pounds published a story in *BBC Wildlife* magazine titled "Disappearing Gold." In protected areas like Monteverde where deforestation was not a factor, Pounds wondered if "the general degrading of our biosphere" might have reached a tipping point, where amphibians began to feel the impacts. Their permeable skin made them particularly vulnerable to environmental disturbances, which had been metastasizing over the last century with the sunrise of the anthropocene—the age of man. These amphibians belonged to two worlds—water and earth—and so found themselves exposed to twice the peril. Offering a perspective that would go on to become as pervasive as it was contentious, Pounds proposed that the disappearing amphibians, like the golden toads, might be "bellwethers ringing a message"—a warning to the rest of us, the first to fall into the cracks of doom, but not the last. Toward the end of the 1990 *BBC Wildlife* article, Pounds suggested that "some pathogen (a virus?) or parasite" might have had a hand in decimating the local populations—but the scale of the disappearances seemed to Pounds beyond the reach of a lone killer: There must have been an accomplice at work.

In 1994, Alan Pounds and Marty Crump reunited to publish a theory informed by this suspicion, for the first time implicating climate change in no uncertain terms. Thanks to the local John Campbell and his highland weather station, Pounds was able to examine twenty years of Monteverde rainfall data, and this local history led him to an anomaly: a climate disturbance associated with the 1986–87 El Niño event, which had caused unusually warm and dry conditions in 1987 and saw the flow of local streams fall to a record low. The data lined up with Marty Crump's own observations during the 1987 season, when she had witnessed the breeding pools of the golden toads dry up in April and May, and only twenty-nine tadpoles had survived.

With these climate disturbances in mind, Pounds and Crump proposed the "climate-linked epidemic hypothesis," suggesting specifically that abnormal climate conditions had made amphibians more vulnerable to a microparasite: a bacterium, a protozoan, or a virus.

Pounds would spend the next five years exploring the hypothesis that the extinctions were entangled with the changing climate. In April 1999, his *Nature* paper, coauthored with Michael Fogden and John Campbell, described "a constellation of demographic changes" in the tropical forests of Monteverde that impacted birds and reptiles in addition to amphibians—changes linked to recent warming. Not only had the dry season lost significant mist frequency over the last twenty years, but the atmospheric warming had raised the altitude of the cloud bank. For highly specialized species like the golden toads, who lived at the top of the mountain already, there was no place left to go when the mist and clouds rose above their reach. And Pounds was still haunted by the forty dead and dying harlequin frogs that he had seen on the banks of the Río Lagarto, victims of the lethal parasitic flies that had grown more deadly with the drier weather.

"Because climate affects host-parasite relationships and amphibians in various ways, it may have set the stage for similar mortality events," Pounds wrote, "including those ascribed to chytrid fungus outbreaks."

By 1999, of course, the warrant was out for chytrid.* The papers from the collaborators at Urbana had identified the theoretical pathogen that had been a part of Pounds's hypotheses. With the pathogen identified, Pounds acknowledged *Batrachochytrium dendrobatidis* to be a likely candidate behind the local disappearances, but he considered it unlikely that this single cause could explain the patterns they were seeing in Monteverde. "At least one other factor," he wrote, must have played a role. Extreme changes in moisture and temperature could raise the probability of local pathogen outbreaks, and the same environmental conditions could weaken the immune systems of the local amphibians, increasing their vulnerability to infection. If the chytrid fungus was the bullet that had killed the golden toad, Pounds was convinced that climate change had pulled the trigger.

At the start of the new millennium, most biologists on the trail of lost frogs agreed that the *Batrachochytrium dendrobatidis* fungus had played a central role in the twentieth-century amphibian annihilations, but

---

* The technically correct terminology should be "the chytrid," as "chytrid" is merely the abbreviation for the entire phylum. However, "chytrid" is commonly used today as a shorthand for *Batrachochytrium dendrobatidis*, and therefore we will occasionally employ it in this manner regardless of the fact that this is admittedly unfair toward the vast and otherwise innocent Phylum Chytridiomycota. For this, we ask Joyce Longcore's forgiveness.

they would continue to debate the mechanisms of the epidemic—where it came from, how it spread, and why it killed.

The beginning of that story goes back nearly two and a half billion years, to a time long before the first frog clambered onto the soil of the earth. Kingdom Fungi, of which chytrid is a part, is one of the oldest and most complex groups of living things on the planet—and fungi are ruthless survivors. Fungi were present in our seas when the sky of nitrogen and ammonia and carbon dioxide burned an alien orange; fungi survived our planet's first apocalypse, when runaway cyanobacteria filled the atmosphere with oxygen and sent themselves spiraling toward extinction. By some accounts, it was a fungus that first crept forth from the cradling sea to brave the unknown future of life above the surface, on the shores of the supercontinent Rodinia.

Around the time of the Cambrian explosion, the earliest chytrid forms began to emerge from the tangled branches of Kingdom Fungi, developing the flagellated zoospores for which they would be recognized, and the pot-shaped structures for which they would be named. When the volcanic eruptions began at the end of the Permian period—poisoning the air and boiling the earth, turning the seas to acid, wiping out nearly three-fourths of the planet's life—many fungi prospered amid this graveyard earth; it may have even been the age when fungi dominated the planet. And when asteroid and impact winter dethroned the dinosaurs, the chytrids were there, blooming in the liquid ashes as though nestled amid a planetary compost heap.

From the crater of the K-T impact crawled the progenitors of our modern frogs and toads, and some may have even then carried chytridiomycetes, unknowing and unbothered by the fungal spores that coexisted harmlessly with them. Chytrid was not born a killer, after all, but became one—through the long, slow marches of mutation and response, of natural selection and evolutionary arms race. In time, a line of chytrids found a benefit in parasitism, exploiting the vulnerabilities of the amphibians they existed alongside. And those amphibians bore them away to dark corners of the world. There, deeply sequestered and cut off from all others of their ancient race—in the jungles and ponds of what would later be named the Korean Peninsula—the frogs developed their own resistances to the parasitic assaults of the chytrids. Terrible as those fungal assailants had grown, the frogs learned to live with them. There, in that secluded world, the jungle found its balance once again.

But between the tangled roots, deep within the murky pools, the chytrid brooded. Its victims had grown, at last, invulnerable to its assaults—but it had lost none of its potency. It needed only to wait, patiently, until a new victim wandered down within the long reach of its grasping arm, helping it to spread beyond its geographic sequestration.

That help would come, in time, from the long-sundered descendants of the mammals that had evolved alongside chytrid throughout the long millennia. *Homo sapiens*, growing in power, had extended their reach to span the globe, now traveling freely across the continents in the blink of a geologic eye—and taking with them all manner of ideas, and tools, and diseases. One of these, it would appear, was chytrid.

The vanguards of chytrid's global crusade rode unnoticed upon the backs of carrier frogs stowed away in shipping crates in the bellies of dark ships, bound for new worlds. In some cases, small fires of infection flamed to life, only to burn out as quickly as they had come—cases of chytrid have been found as early as 1915 in Southern California and 1926 in Baja, though the disease appeared to die out before spreading in any meaningful way. In other instances, the fires of chytrid did manage to establish themselves among the local populations, burning low like coals, but still localized and relatively insignificant in effect; in time, these strains of chytrid would grow distinct enough from their sources to be seen as unique lineages: *Bd*ASIA-1 in Korea, *Bd*BRAZIL in South America, and *Bd*CAPE in South Africa.

And that might have been the end of this story, with no amphibian crisis to speak of—no midnight disappearances, no empty pools and silent springs at the tops of barren mountains—had it not been for

*Bd*GPL: the Global Panzootic Lineage. Arising from the meeting of two strains of genetically isolated populations, *Bd*GPL was the snake-eyed roll on the dice of fate—a hypervirulent combination of an overzealous parasite, an evolutionary error that would result not in equilibrium but annihilation. All it needed was a method to spread.

This method would come, in part,* in the form of a British scientist named Lancelot Hogben. In the 1930s, while living in South Africa, Hogben began to collect African clawed frogs (*Xenopus laevis*), which he found a useful and easy-to-acquire subject in his hormone studies. It was in the midst of one of these studies that he injected *Xenopus* with extracts from an ox's pituitary gland, and watched with interest as the frog soon began laying eggs. In an unusual leap, Hogben considered that the urine of pregnant women contained hormones also brewed in the pituitary, and realized that the frogs could perhaps be used as a living pregnancy test. Only three years after his arrival, Hogben fled South Africa as a result of his outspoken ideas against apartheid (he and his wife were once reported to have hidden two locals fleeing a lynch mob in the trunk of their car), but he resumed his experiments in Britain—with a colony of *Xenopus* in tow. The "Hogben Test," as it would come to be known, soon proved simple and effective: Inject a woman's urine under the skin of a female *Xenopus* and, in a positive case, the frog would produce a cluster of eggs within twelve hours. The process was cheap, fast, and more accurate than most existing methods. All it required were frogs—a lot of them.

As the Hogben test became widely adopted in the 1940s, the export of *Xenopus* from Africa reached a fever pitch. A contemporaneous description of the test reported that "animal dealers seem to have no

---

* The global spread of *Bd*GPL was not accomplished through a single instance or means of movement, but through many distinct cases over long years, each adding its weight to what eventually became a critical mass of distribution. Hogben's unintentional complicity has, at times, been overstated as a contributor. However, the global trade of *Xenopus* certainly played a part, and stands as a clear example of the types of actions that conspired unknowingly to accomplish chytrid's global proliferation.

difficulty in catching as many as are required" and that "supplies seem to be unlimited and export unrestricted." Over the ensuing years, thousands of frogs were exported across the globe, with some locations receiving roughly a hundred frogs a month. Some reports state that the crates the frogs were shipped in were "quite tricky to open without allowing a frog stampede to occur," and stories abound of fridge doors failing to shut properly, and lab workers subsequently greeted by "an army of wanderlusting frogs." Inevitably, frogs managed to escape, and invasive populations were established in the wild. By 1970, as a result of its unusual value, *Xenopus laevis* had become the world's most widely distributed amphibian.

And carried on the bellies of the African clawed frogs was *Bd*GPL. Well adapted to and largely unbothered by the pathogen themselves, *Xenopus* carried the disease unnoticed, like a dark secret, into the unspoiled ponds of the world—a misguided vengeance, perhaps, for their mistreatment at the hands of humanity.

Introduced by this global trade into previously unexposed populations, *Bd*GPL unleashed itself upon the world's unsuspecting amphibians, spreading with unprecedented swiftness, exacting an incomprehensible toll, and charting an inescapable course for destruction. By 1978, it had reached a port in Brisbane, and from there it spread rapidly through Australia, falling upon the globally unique gastric-brooding frog, the familiar southern day frog, the sharp-snouted day frog of the high pools, and dragging them down to the depths of extinction. Prior even to this, in the early 1970s, it had arrived in Mexico. Sometime in the 1980s, it struck Guatemala, where it rampaged through the Sierra de las Minas mountain range, decimating populations of the Jalapa toad, Barber's sheep frog, and Bocourt's tree frog.

But even by the early 2000s, many of the questions around the chytrid's mechanics and movements remained unanswered. Had chytrid reached the golden toads on their high mountain? Were Monteverde's precious gems another victim of the pathogen's roiling tidal wave?

"There is no evidence," Alan Pounds had written in 1999, "that a single outbreak, spreading in a wavelike fashion, has caused all the declines in lower Central America."

As the new millennium began, Karen Lips set out to find that evidence.

El Copé, Panama

More convinced than ever of chytrid's north-to-south spread across the Central American isthmus, Karen Lips unfurled a map and traced her finger in a line from Las Tablas, to Fortuna, and beyond. There on the highline of Panama's Central Cordillera, not far from the Gulf of Parita, she found the place she thought was likely to become the next haunt of the killer: El Copé National Park. It was the right temperature, the right altitude, and dead in the path of what she believed to be an epidemic wave.

Lips wrapped up her classes' final exams and then lit out to establish her new site at El Copé—where she was sure she would witness the arrival of *Batrachochytrium dendrobatidis*. And if she did, she would be ready for it. Shortly after the Urbana meeting, Lips had paid a visit to Joyce Longcore, the uncontested expert on the killer chytrid. As a snowstorm buried the nearby fields in deep drifts, the two scientists had hunkered down in Longcore's small lab at the University of Maine. Over canned preserves and cranberries, Longcore had revealed her carefully refined technique for growing their killer in culture.* Armed with this new knowledge, Lips and her team swabbed the frogs at El Copé; they tested negative for *Bd*. All that was left to do was to settle in and wait.

"We spent six years," Lips recalled, "waiting for something to happen."

---

* In the years that followed, Joyce Longcore would leave behind her humble laboratory to travel the world, teaching scientists around the globe what she taught Karen Lips. Without her efforts, the penultimate chapter of this book would have unfolded very differently.

2004—the year that Karen Lips would finally catch the killer in the act at El Copé—would prove to be an important year for the mystery of disappearing frogs. That February, Alan Pounds and Robert Puschendorf published an article in *Nature* titled "Clouded Futures." Along with documenting the ways that global warming was existentially changing the ranges and abundance of plant and animal species, they also specified how climate and chytrid might have colluded in the decimation of these species—*Batrachochytrium dendrobatidis* was believed to thrive under cool and moist conditions, and recent experiments had demonstrated that as frogs basked in the sun to elevate their body temperatures, they could rid themselves of the chytrid fungus. "Both increased cloud cover and unusually dry weather might hamper these defenses," wrote Pounds. The dry conditions recorded in Monteverde during the 1986–87 El Niño might have limited moisture and forced the toads to remain in cooler, damper places—lending deadly strength to an outbreak of *Batrachochytrium dendrobatidis*.

That same year, *Nature* published a major paper titled "Extinction Risk from Climate Change." Citing the golden toad as a species-level extinction in which climate change had been implicated, the authors projected species distributions for climate scenarios to predict extinction risk; their results indicated that with mid-level climate warming by 2050, up to 37 percent of the species in their samples would be "committed to extinction." The authors concluded that anthropogenic climate warming "is likely to be the greatest threat in many if not most regions." Echoing Pounds's concern for synergistic factors, they also wrote that "many of the most severe impacts of climate change are likely to stem from interactions between threats."

And in September, Karen Lips received the news she'd been awaiting for the past six years: The killer had finally arrived.

"I immediately felt panic," Lips would reflect later. "It was happening."

The first infected frog at El Copé, in the cloud forests of highland Panama, was discovered on September 23, 2004. From then on, the nightly chorus was tinged with discord, a note of mourning underneath the ancient music—cries of sorrow echoing in the dark. The killer had reached the frogs; it was among them. There was nothing they could do now to escape it. On October 4, the first dead amphibian tested positive for *Batrachochytrium dendrobatidis*. Over the next three and a half months, the team would identify between one and nineteen mortalities every day.

"It was just heartbreaking," says Forrest Brem, one of Lips's graduate students who was on the ground in El Copé. "I'd heard people describe El Copé as 'Disneyland for Herpetologists.' There were almost as many species in that little two-by-two-kilometer area as there are in the entire United States. And then to see these animals disappear . . . not even dead yet—still alive, but unable to do anything because chytrid screws up their ion balance."

To die by chytridiomycosis was not an enviable way to go. The fungal zoospores reached their victims either through direct individual transmission, passed from frog to frog among amphibian communities; or lay in wait in the shadowed soil, crouched upon the wet darkness of rocks or swimming freely through the water with their flagellated tails. No matter the manner in which the victim was infected, the outcome all too often looked the same. Embedding in the skin of the infected amphibian, the chytrid zoospores shifted into a reproductive mode and began duplicating, slowly overwhelming their host as they fed on the keratin and other proteins in the frog's skin—the most sensitive and critical of amphibian organs. As the skin peeled away or sloughed off entirely, the frog's life-support systems began to fail, manifesting in lethargy and disorientation. With the skin sufficiently impacted, a fatal imbalance of water and electrolytes accrued; if the infected frog didn't first become the meal of a passing predator, it died from a heart attack.

Trekking through the forest trails, Brem and his colleagues faced these sickening outcomes of the killer's work—twenty years later, he recalls the heap of broken images from those days: of lifting a still-living frog out of the water to find a crab hanging onto its leg and eating it alive; of kneeling to inspect a Panamanian golden frog, and finding only the shell of its skin left, the rest of it hollowed out by scavenging ants.

As the new year dawned, the incidence of corpses finally began to lessen. But even that was not a blessing; they were finding fewer dead frogs because there were fewer and fewer frogs left in the forest. By February 2005, when the forest had largely gone silent, the team had documented 346 deceased frogs. *Batrachochytrium dendrobatidis* had wiped out thirty species from El Copé, reducing the overall amphibian population by more than 75 percent. It was the outcome that Karen Lips and her team had been waiting for and dreading.

In March 2008, Karen Lips and three coauthors published their research to support the hypothesis that *Batrachochytrium dendrobatidis* was spreading in a wave across Central and South America, in what they deemed a "spatiotemporal spread." In addition to proposing the spatiotemporal spread as the primary mechanism for the dispersal of the chytrid fungus, Lips's 2008 paper was also a determined rebuttal of Pounds's climate-linked epidemic hypothesis. She stated that her analyses found "no evidence to support the hypothesis that climate change has been driving outbreaks of amphibian chytridiomycosis, as has been posited in the climate-linked epidemic hypothesis" and that "the available data simply do not support the hypothesis that climate change has driven the spread of *Bd* in our study area."

Lips's theory sent ripples into the ongoing debate. In 2009, Marty Crump coauthored a book titled *Extinction in Our Times: Global Amphibian Decline*, and dissected the chytrid vs. climate debate; the authors acknowledged that the warming climate would predispose some animals to infection, but argued that "the amphibian chytrid's invasion of naïve populations" was "the best explanation for the enigmatic, worldwide declines."

To Alan Pounds, this was an oversimplification of the argument, and he responded in a 2009 note in *Nature*, coauthored by Karen

Masters: "This is like attributing a car crash to excessive speed and deciding that other contributing issues, such as alcohol consumption, need not be considered."* The chytrid fungus had almost certainly played a major role in the amphibian extinctions, they agreed, and would likely go on to other reigns of terror; but it was not the only threat to frogs, and frogs were not the only creatures that should be concerned. "The interacting changes threaten many life forms," Pounds and Masters wrote.

Although they were all working toward the same goal—understanding the amphibian extinctions they had witnessed so they might prevent the recapitulation of calamity—the biologists trying to identify and confront the killers found themselves increasingly at odds with one another. In a world in which it was maddeningly difficult to raise money and awareness for research around the amphibian crisis, the biologists in both camps—chytrid and climate change—were competing over a finite reservoir of funding and support, as had been the case for some time. When Lee Berger presented on the Australian discovery of the chytrid infection in the country's frogs at the Third World Congress of Herpetology in 1997, she found her session sparsely attended. "And honestly," she says, "I don't think anyone believed it." Another Australian who played an instrumental role in bringing attention to the early disappearances declined to be interviewed, stating simply: "Getting the scientific world to notice the fact of the sudden decline of frogs in the 1970s was an awful experience." It wasn't only research funding that the biologists were competing for, but attention from the general public. The theories they were developing were complex: a pathogen outbreak that spread across continents, the mechanics of climate dynamics in natural ecosystems. Paradoxes had arisen, contesting

---

* They also noted that the spatiotemporal spread was still being evaluated: The Costa Rican biologist Federico Bolaños had suggested that the data could be impacted by biased sampling and incomplete perspectives on the regional amphibian declines.

theories, data-driven observations that were not easy for non-scientists to understand. It began to feel like a zero-sum game.

"I would have to say that it was a fairly painful period, to be honest," says Karen Masters. "There was a decade where there were a lot of fiery angry emails and conversations. There was a lot at stake."

For the biologists like Masters who were still living and working in Monteverde, it was not something that was easy to forget; the cloud forests that had once been full of frog calls were quiet now, and the golden toad was still missing. When Karen Lips and her coauthors published their findings on the spatiotemporal spread in 2008, they included a map of Central America to visualize the wave of the deadly chytrid fungus. The map included dates and locations for the sites where the chytrid had been implicated: 2004, El Copé; 1996, Fortuna; 1993, Las Tablas. Each location marked the hypothesized leading edge of the wave of *Batrachochytrium dendrobatidis*, with rates of spread denoted in between: thirty-three kilometers each year between Las Tablas and Fortuna, fifty-eight kilometers each year between Fortuna and El Copé. But in the upper left-hand corner of the map, to the northwest, there was a lonely circle balanced on the Tilarán cordillera of Costa Rica, the first Date of Decline denoted on the map: Monteverde, and the golden toad.

Monteverde was the ghost story that had started the alarm bells ringing, the first piece of the puzzle that Karen Lips would assemble to begin to map the spread. When the first frogs had disappeared from Las Tablas, it was the golden toad that Lips had thought about; and when she began to chart a pattern, she had calculated a rate of spread between Monteverde and Las Tablas at forty-two kilometers per year—a rate consistent with the outbreak of a pathogen. That trajectory led her to Fortuna, and on to El Copé. But even in light of everything they'd learned, the mysteries around the golden toad remained.

"Although it is widely assumed that the decline of amphibians in 1987 at Monteverde Cloud Forest Reserve, Costa Rica, was the result of an outbreak of *Bd*, direct evidence of such does not exist," Lips acknowledged in her 2008 paper. As part of her research, Lips had examined museum specimens of preserved Monteverde frogs collected between 1979 and 1984, prior to the population declines that began in 1987; none of the sixty-four specimens was infected with *Batrachochytrium dendrobatidis*. A more recent study published in 2013 tested fifteen golden toad specimens collected in April 1982; all skin swabs tested negative for the fungal infection. In the same 2008 paper, Lips referenced a positive record of *Bd* in Monteverde in 2003, indicating that *Bd* had become established in the area. For Lips, these findings served to support her hypothesis that the chytrid fungus was an invasive introduction, not native to Monteverde before the advent of the declines. For others, the matter has proven more uncertain.

San Pedro, Costa Rica

The University of Costa Rica in San Pedro hadn't changed much since Pri had last seen it as an undergraduate eleven years before. When she and I returned in 2018, we took a taxi to the outskirts of the campus, and as we wandered the paths winding among the buildings, she pointed out the new trees growing on the lawns, or the students who looked younger than either of us remembered from our college days. But the old concrete buildings were in the same places that she'd left them; it didn't take her long to find the Biology building with its deep hallways, and her old Ecology professor—Federico Bolaños—behind a great white beard in a dark lab room looking out over the campus. At the end of the 1990s and in the early 2000s, Bolaños had boots on the ground as one of the most knowledgeable herpetologists in the country, describing new salamanders, assessing conservation impacts, and studying amphibian declines. He remembered Pri from his classes a decade before—calling her nickname *"Pucca!"* in his deep voice as we walked inside. As Pri and her former professor shared old stories, I carefully unpacked my recorder and set it on the desk between us. Bolaños looked at it distrustfully from underneath white eyebrows. He knew why we had made this particular pilgrimage, and the questions we had brought for him, many missing answers.

"I do not really believe," he began, "that we have enough information to understand what happened with the golden toad."

The chytrid fungus had been incriminated in the mass mortalities of other frogs in the tropics and beyond; but in the case of the golden toad, there was no decisive evidence that the pathogen had been at work, particularly because no bodies had been found to test for chytrid before the golden toads had vanished.

By the time that I reached Costa Rica, almost thirty years after the golden toad's disappearance, the long debate about the species still persisted. At a certain point, most of the investigators on its trail acknowledged that the majority of the evidence was circumstantial. With the evidence and data from the other crime scenes, we can place the killer on that mountain at that time, but may never be able to convict it.

"The fungi was present in Monteverde at the same time the golden toad declined," Federico Bolaños said to Pri and me when we visited him at the University of Costa Rica. "However, there is no proof that a golden toad had the fungi." He had read the papers and considered the theories, balancing data with chisme,* and he had made his peace with the uncertainty. "There is really good information that shows how the climate has changed in a place like Monteverde," he said, "and there is also information that shows that those diseases have been present in Monteverde since the declines. We're not certain. And I prefer to say that we need more information to be sure about what happened. But I believe that the combination of the factors could be the reason."

When Pri and I spoke with him, Bolaños implicated another suspect: time. "Maybe we don't know the real story of the golden toad,"

---

* Chisme: Rumor or gossip.

he acknowledged. "But with Monteverde and the golden toad, it's the perfect case to say that the golden toad was destined for extinction." The species was so specialized, and it had made its home at the top of a mountain, as high as it could climb. Even small changes in that temperature and climate would have had a cascading impact. Bolaños told us that he believed its extinction had been inevitable, though he knew that many people wouldn't like to hear it. "If it is true that temperatures are increasing, then there was no place for them to go on that mountain. They had nowhere to run. I really think that the golden toad is an example of a species that was for sure going to become extinct."

When our conversation drew to a close, I turned the digital recorder off and started packing up. We thanked Bolaños for his time and histories and wandered back into the sun. I was lost somewhere in my own thoughts; guessing how the golden toads might have felt in the last days on their lonely mountain—site of wonders, doomed from the beginning. In the end, could the very principle that had made them so remarkable have been the instrument that killed them? Was it all inevitable—their endemism, their annihilation?

There are two ways that a species might become endemic. In some cases, the species would have once possessed a much wider geographic range, making its home across far land masses and at different elevations. Over slow time, as the mountains rose and the earth shifted underneath them, they would be separated from one another or die out until they were left in a final stronghold, the very last place that they existed. In this case, we might look at the mechanics of endemism as a dark mirror to the proliferation of disease, a clock running in reverse. In the same way that the chytrid fungus grew and spread from its initial point of origin and evolution, perhaps the golden toads retreated, dying off in farther ranges one by one, until across long ages they found themselves alone on the last mountain, waiting in the cold and dark for an unknown killer to finally reach out and touch them. And here we might see Monteverde as the last outpost of a once-expansive kingdom, beautiful and sad, holding memories in its cupped hands that would soon be gone forever.

But it is more likely that the golden toads evolved on the backbone of the cordillera de Tilarán—in dark ages when the hills were young—and that those forests were the only homes they ever knew. When the troubled waters and emerging mountains rose to sunder the wild highlands of Central America from the north and south, the ancestors of the golden toads might have climbed into the misty elevations or ridden the rising mountain ranges to find dark hollows in the secret forests, depositing the coming generations in the sweet waters among the roots of ancient trees; and finally the deep woodlands were alight with a new endemic gold, El Dorado, the magnificent. In that case, we might instead see Monteverde as a cradle of life, where all the odd conditions were in perfect harmony to invite the strange and beautiful species into existence—the one and only home of the golden toads, the fountain of creation from which they had arisen. And if that is true, then it means that they ended where they began, in the forests of their ancestors, listening to the same familiar wind and feeling the same rain sent down from high above.

Those were the forests that Federico Bolaños had spoken of—"There was no place for them to go on that mountain"—and the forests that Jay Savage and Norm Scott and Jerry James had cut a trail through on their pilgrimage into the highlands; they were the forests that Wolf Guindon had explored in his homesteading days, and the forests that other earlier settlers unknown to us had called home long before. They were the forests that Ricky Guindon had wandered as a child with his father, and his father had given him a name, and Ricky had passed that name along to us: Brillante, last refuge of the golden toads. The time had come for us to knock upon the door of those enchanted forests. The time had come for us to walk into the past.

CHAPTER SIX
INTO BRILLANTE

Monteverde, Costa Rica

In a silver mist that moved like a ghost under a sheet, the town was still sleeping when we rose in the blue dark of early morning to hike into Brillante. From the porch of the old cabin, we could look down across the rolling hills of the Pacific slope to see the glitter of the Nicoya gulf in the distance, where star- and moonlight wavered eerily on the rolling surface. That same light glinted on the sheet-metal roofs of houses on the hill below, rusted in places, peeking in and out of tall grasses and bamboo.

The smell of coffee was drifting out from the open window, and Pri—still half-asleep—followed the steam from her mug to sit with me and Kyle on the porch. We'd been up for the last hour looking at the maps that Pri and I had drawn out from the Civil Registry in San José, tinkering with the cameras we would carry through the highland forests, and talking nervously about the expedition we were soon to undertake. Across her cup of coffee, Pri eyed the mound of gear that I'd assigned to Kyle doubtfully.

"You're making your brother carry all that stuff?" she asked.

Kyle looked up with a careful smile. "You guys keep handling the Spanish and I'll carry anything you want," he said.

He had flown in to help us film Brillante for the documentary I had promised my thesis committee, called back from his own investigations of the golden toad's suspected killer to walk with us through its graveyard. I was glad that he was here. It had been a long time since we had been together, and it felt like we had wandered into the old days again, back when we had gone off looking for crocodiles in the Everglades or red wolves in North Carolina's coastal swamps. It was a reunion that had been a long time coming, and it felt right to have my family—new and old—come together for the journey. If there were certain rituals required to gain entry into hidden places like Brillante, it felt like we were meeting their conditions.

Pri and I had picked him up in San José a few days earlier, and on the long drive up to Monteverde, Kyle had begun to tell me what he'd learned so far about the chytrid fungus and the lost frogs of the world. The early disappearances in Australia, the dead toads in the labs in Washington, the apocalyptic spread across the tropics—his information, combined with what I'd learned from Pri and the other stories that I'd heard of amphibian declines in Costa Rica, began to weave a disconcerting tapestry. But the troubling fact remained: With no dead golden toads to study, there was no way to be certain about their killer—or their fate.

That mystery made Brillante feel all the more important. That forest was the place the golden toads had lived and died, and the site of their last confirmed appearance. For a long time now, I had wondered if their old pools and hollows concealed the clues to their annihilation. And another thought had begun to gnaw at me more recently: If nobody had found their bodies, could that mean the golden toads might still survive?

The night's last moths descended to worship the pale glow of the false moon when Pri turned the porch light on and began to get dressed for the trail: tall fútbol socks, an old raincoat, dark colors to keep the tábanos* away. Kyle had pulled out his *Field Guide to the Birds of Costa Rica* and was flipping through it silently. Watching the two people I had drawn into the orbit of my obsession, I began to hear the soft feet of uncertainty creeping up to consider our arrangements. This pursuit

---

* Tábanos: Horseflies.

that I had talked them into—the search for the golden toad, for El Dorado—was not a trial I had thought too seriously about. Only here, in the shadow of the odyssey's proper start, did I consider what the quest might cost. I had heard that curses often fell on those who went to wake the dead.

But we had come too far to turn back now—drifted too close to the pull of the mystery. After Ricky Guindon had given us the name we sought, Pri had gotten us permission from the Reserve to walk the old trail that wound through the breeding grounds of the golden toads. They told her that the trail was in disrepair: A couple of times a year, a ranger from the Reserve walked the thin line to cut away the encroaching forest—preserving the way for any research projects that might require it, or out of respect for the memory of the dead—but it had been closed to the public for more than thirty years, erased from all the maps.

When we turned the lights off one by one, the night overtook the cabin again, claiming what belonged to it. The three of us walked to the jeep and climbed in, and when I started the engine the headlights cast dim beams into a forest rising up like a great wall, sending night creatures scattering for shelter. We rolled down the steep driveway to the bottom of the hill and turned left at the cheese factory, crossing the stream and driving up toward the Reserve. We squinted forward into the circle of illumination thrown out in front of us by the headlights, watching for the deadly curves and the places where the road had crumbled in the landslides that had scarred the mountain just a few months back. By the time we made our way above the abandoned pastures to the closed gate at the foot of the Reserve, the sun had started to rise in the gray morning, but it would be a while yet before its light found a way to penetrate the forest.

We parked in the empty lot and climbed out to walk around the gate, which wouldn't open for another hour at the earliest. Later, the Reserve would be full of tourists with big cameras and groups of children on school trips, but for now it was quiet except for the distant calls of birds higher in the forest, and the flap of wings that rained water down from green canopies. Karen Masters was waiting for us at the head of the wide trail running from the entrance up to the ventana on the Continental Divide. Her silver hair was wet and her raincoat was glimmering from the moisture in the air. She'd driven down from her house up on the hill that morning, though she could have gotten here by other paths if she'd desired. She knew the unsanctioned trails that cut down through the canyons below Cerro Amigos and into the Reserve, and the rangers had discovered her from time to time in places she wasn't meant to be.

It had been eleven months since Pri and I had last seen Karen, and we were happy to reunite. She asked Kyle and me how my parents were and how Pri was liking Arizona, and we asked about how life on the green mountain had been treating her. While Pri and Karen reminisced about the old days in Monteverde, Kyle and I wandered off to the little signpost by the trailhead that displayed a map of the Reserve. The most common trails were marked in red or purple lines—El Camino, Sendero Bosque Nuboso, Sendero Roble—but there were others that had been removed from the new maps, leaving empty spaces: swampy Pantanoso with its decaying boardwalk, the upper section of Chomogo that climbed into the clouds, and of course the old Sendero Brillante, a portal into history. I realized that before long we would walk beyond the edges of the map, and I wondered if it hurt to disappear.

Because the forest was swallowing the trail and the terrain was treacherous, the Reserve was sending two rangers along with us as guides—like time-travelers with a map into an inauspicious past. They introduced themselves as Juan and Alvaro—Pri recognized Alvaro from his night job; he was a bouncer at Bar Amigos—and they winked at Karen, pretending they had never caught her on the closed trails that run in and out of the Reserve. While they talked—about the condition of the trail, about the chances we might finally get rain—the last member of our party walked up the dirt road by himself.

From a distance, Eladio Cruz looked old and small in the shadow of the imposing forest. He had walked up in the early morning through Monteverde from his home above the old Sapo hotel in Cerro Plano, five kilometers along a footpath on the edge of the valley, over paved roads and dirt roads and across a wet tree over the river. He, too, knew the old ways through the barrios and bosques. The rangers greeted Don Eladio* like an old amigo—a friend of their fathers, maybe—because everybody knew Eladio: His presence seemed to give our task a blessing, the consecration of a pilgrimage that might be holy or unholy.

In the days that followed our conversation with Ricky and Hazel at the Guindon house, I had grown more and more convinced that the time had come to arrange a reunion with Eladio. We had our maps, and we knew the name of the trail that we were seeking, but we were still in need of a guide to lead us through the tangles of the past. Eladio knew those forests well, from his early years spent clearing pasture for Wolf and, later, helping Wolf and Powell draw the boundaries for the Reserve. And Eladio was on my mind for another reason. I had not forgotten what my old friend Moncho had told me, three years earlier when I had first inquired about the golden toads. "It is depending on who you ask," he'd said as we'd sat together in the biological station. "But there is one person who will say that he saw it later, at another place." That person was Eladio Cruz. His story was one that I still hoped to hear, if he judged that I was worthy of the revelation.

When we'd met him a few days earlier at the Café Orquídeas in Santa Elena, Eladio had come in looking shy and humble, and he sat with us for a little while over coffee while we told him about our plans to walk into Brillante. Pri had known him from her days in Monteverde, and he remembered me a little from our scramble into Peñas Blancas years before, so he agreed to join us on the journey to explore the old trails where, he told us, it had once been common to see a thousand

---

* In Costa Rica, the prefix "Don" is a sign of respect for an elder or honored person.

golden toads. We'd parted ways, making plans to meet at the Reserve, and he had seemed a little sad. For him, it would be like going back into the home of an old friend who had died. I hadn't asked about the rumor of his later sighting then—I had the feeling that it was the kind of question that could only be asked once, and, if it wasn't answered then, it never would be.

Reunited now, we set off up the old camino through the forest, toward the ventana and the hidden entrance to Brillante. The camino was the wide horse trail that the old settlers had used when they crossed over the Continental Divide to drop down into Peñas Blancas, back before the Cloud Forest Preserve even existed. Eladio had taken that trail many times to reach his little farm and homestead in the valley; but now the homesteads that once dotted the canyon slopes along the river were overgrown, abandoned, and forgotten. The ruins of the remaining structures were almost mythical—the Alemán Shelter, Eladio's Refuge—way stations for the rangers or the biologists who still gambled on odd wanderings across the wild stony river and along the misty ridges.

We walked the wide camino at a brisk pace, crossing the old quebrada running down from the highlands and dipping beneath the elephant ears that lined the steep path climbing to the ridge. Kyle trailed behind the bulk of our group, weighed down by the camera gear that he was carrying, and stopping every now and then to record a sweeping panorama through the treefall gaps, where the land fell away toward the sea. These trails were not familiar to him; every vista brought a new perspective, every bird call a new language to decode. My gaze had grown narrower through the years; he was seeing the forest for the first time, in a way I never would again.

I spoke a little with Eladio as we went along, but my Spanish wasn't sharp enough for the questions that I longed to ask him: *Do you still feel the spirits of the golden toads in this forest? If you could speak with them, what would you say? If they could answer, what would you fear to hear from them?*

INTO BRILLANTE

We squinted through the fog and ducked down beneath the bending trees to shelter from the blowing mist on the windy ridge, ending up at last on the wooden platform at the end of the trail. We stopped there for a few minutes to look down on the Pacific and Atlantic slopes falling off on either side, catching our breath from the brisk hike and waiting for Kyle to catch up with his heavy tripod. I took a few pictures of the impenetrable forest that tangled out along the ridge. But soon there was nothing left to do, no more errands to delay our incursion. Sensing this, the wind blew a slow gust up the trail behind us, sending the woods that lay ahead into a slow and solemn dance, a ritual.

"Está allá," Eladio said softly, nodding to the veiled entrance of the mythic trail. "Brillante."

It was overgrown and nearly imperceptible in the shadows of the forest—a small gap in the warding branches, the echo of a path that had once been famous. Now there were few left to remember it; we had brought one of them with us.

Eladio's eyes were dark, withholding judgment, or maybe only speaking in a language that I didn't understand. Juan and Alvaro led the way with Karen, pulling aside the hanging vines and tall bamboo, and Kyle, Pri, and I followed them into the past. Only Eladio hesitated before pushing his way in, as if he knew the peril. Then, after a moment, he, too, walked into Brillante.

Monteverde, Costa Rica

We were not the first to go looking for the golden toads. When Frank Hensley had taken his last photo of the lonely endling[*] on the Brillante ridge in 1989, he had no reason to suspect that he was witnessing the end of something. After his professor Marty Crump returned to Florida, Hensley spent the next four weeks hiking up to Brillante and walking

---

[*] Endling: The last known individual of a species.

the circuit between the known breeding pools. Day after day he measured the pools, took the water temperature, recorded the air temperature, checked the rain gauge, searched the root hollows: no toads, no toads, no toads. Mystified, he widened the boundaries of his search, hiking out along the ridgeline with Wolf Guindon to explore Cerro Ojo de Agua, where Wolf had seen nine toads in 1988. They returned exhausted* and empty-handed. By the middle of June it was pouring rain. The breeding pools in the Brillante area were flooded, streams of water running down the mountain. Even if the toads had reemerged, their eggs would have washed away in the torrent. Reluctantly, Hensley abandoned the search after two months in Monteverde, hoping to hear news someday that the toads had returned to the mountain.

Wolf Guindon had been on the hunt since early March, even before the first rains had started; before dawn, he would start off from the Guindon farm with a handful of candies in his pocket, carrying his big twenty-six-inch machete and howling at the moon. He would fight his way through deep mud to cross the boundary trail and the old refugio, following streams of water to peer into the dark depressions around the roots of gnarled trees. But he'd found the pools empty, one by one. For the second year in a row, in all the years since he'd first cut his way onto the high Brillante ridge, the golden toads were nowhere to be found.

Wolf continued to patrol the outer boundaries of the Reserve that year, in places like El Valle along his old tapir trail to Arenal where he had seen golden toads in the early years. When the rains ended and the windy-misty season began, Wolf, too, gave up the search and returned to his other responsibilities; he ran his family's dairy farm, attended the community Quaker meetings, and cut new trails down the valley into Peñas Blancas. But when the rains came back over the mountain, he climbed up into the highlands to look into the old pools again, hoping to discover gold.

---

* "I was exhausted," Frank Hensley clarifies, "but Wolf was tireless."

"One more trip—una más—and it's dedicated to Dr. Alan Pounds and his hopping woodland subjects," Wolf recorded high on the Brillante trail in June 1990, his hat pushed back so he might feel the last light of the fading evening on his face. "It's been a long day because Alan and I sure looked for them toads. We shook out every pool. This will be the last official trip this year to see if we might find the toads. It was basically dry on top, although there were places where the pools were still full of water. It looked pretty good, but there was no sign of eggs or tadpoles."

Another fruitless search, another temple empty. Descending from the mossy elfin forest empty-handed, Wolf cut cross-country and stopped at Windy Corner long enough to watch the mist drifting in the hills above the San Luis Valley far below. That night, in the shadow of the tall oak on the Guindon farm, he scribbled off a letter to Marty Crump at her new field site in Argentina: The weather conditions had been good this season, he wrote, but the golden toads hadn't reemerged. Then he walked up to Alan Pounds's property to tell him that the golden toads on the Brillante ridge had disappeared for good.

"I didn't really get interested in golden toads until they were gone," Pounds remembered later. When he began to seriously investigate the Monteverde amphibian decline, Pounds considered the golden toad to be a piece of a larger puzzle—so he, too, joined the search. Earlier that year, he and Giovanni Bello had received funding from Stanford University's Center for Conservation Biology to survey the known breeding areas of the golden toads. For four months, in concert with their broader survey of the amphibians of Monteverde, they walked the windswept trail through the Brillante sector, peering into dark root hollows and pools brimming with water, but their pilgrimages came up empty. When the funding expired, they kept looking. But with every passing year, they knew their odds diminished. If the large aggregations that Crump had observed in 1987 had indeed retreated to their underground hideouts to wait out whatever calamity had befallen them, their clocks were winding down. Without new eggs tucked faithfully into the shallow pools on the lonely mountain, there could be no new generation, no legacy, no hope.

A few days before we ventured into the elfin forests of the Brillante saddle, I talked with Alan Pounds at the old casona at the foot of the Reserve, while a coati wandered in the undergrowth nearby, looking for a meal. That day, twenty-nine years after Frank Hensley's final sighting, he told me he was still looking. "I haven't given up hope entirely."

As the 1990s drew to a close, that hope was like a small flame in a high wind. Even if the last toads observed in Brillante had initially survived, their time had run out. Hope for the species dwindled to the hypothesis that they were alive and breeding in other areas, maybe places where they had never been observed before. After more than a decade of no sightings, the International Union for the Conservation of Nature formally revised the status of the golden toad to "Extinct." It was one of the first amphibians in the world to receive the designation.

The forests creeping up and over the Continental Divide in Monteverde were deep and treacherous, and the naturally secretive golden toads—if they had managed somehow to escape extinction—were not looking to be found. Even for the people who still remembered them, it wasn't easy to chase the rains season after season looking for a ghost. But in the thirty years since their disappearance, hopeful locals and ambitious foreigners have often found the call too bewitching to resist.

As recently as 2014, a team of biologists from the United States and Mexico launched a crowdfunding campaign that they described as "the first coordinated effort to find the toad during the proper time." They selected thirteen different field sites for investigation, including original golden toad localities and new sites based on elevation, rainfall, and habitat. Several areas were outside the protected boundaries of the reserves, requiring coordination with local landowners and access to remote locations. Despite acknowledging the challenges of the expedition (which included access logistics, weather conditions, breeding behavior, and "extinction"), herpetologist Chris Grünwald sounded confident in an interview with the *Tico Times*: "It's still out there. I am convinced, and I am pretty positive we can find it." The expedition never occurred. The crowdfunding campaign outlined $22,000 of proposed costs and fell more than $13,000 short of its $15,000 funding goal. The high cerros and swamps that were marked with X's remained unexplored; the mystery endured.

In every story of the golden toad I heard, in every myth and dream and recollection, Brillante was at the heart of all of them. It was the site of first discovery; then the castle of a golden age; then a graveyard, full of spirits. When the golden toads disappeared, the trail was closed and erased from the tourists' maps, like a curtain falling on a theater, last performance, closing night. But it was still up there, burdened with its history, waiting for the long story of the golden toad to find its way through the old, familiar forests once again.

Monteverde, Costa Rica

The grasping branches closed behind us as we descended from the dripping platform of the ventana and onto the Brillante trail. Hidden from the sun by the arms of twisted trees and the drifting mist washing over the Divide from the wild lands of Peñas Blancas, we paused below the wooden platform to wait until our eyes had adjusted to the changing light. Up ahead, we could make out the drifting forms of Eladio and the rangers and an old signpost with gold letters peeking through the moss—"Do Not Enter"—but by then its last warning was too late. We moved on into the haunted forests that lay ahead of us.

The trail was a bony arm pointing toward a prophecy, but all the ground on either side was covered in green leaves, the shrunken canopy of a pygmy forest of unknown depth. It gave you the feeling that if you were to drop something, it would vanish. As we made our way farther out along the ridge, I looked for signs of the old refugio—the collapsing shack where Wolf Guindon had spent nights waiting out the unexpected storms of his settling years, where Jay Savage and Norm Scott had lain awake listening to the golden toads grasp against the slick glass of their collecting jars—but I couldn't even find the carcass of it; the hungry years had drawn it down into the undergrowth, whatever had remained.

In the lower elevations, the big trees were the monarchs of the landscape—thick buttress roots reaching deep into the earth, long hands grasping at the sweet, warm center—but here on the ridge, the trees were thin and tangled, twisted into crooked bodies. Arthritic limbs grasped from the boundaries of the trail, and here and there one of the rangers would cut one down with his machete, or break a branch by hand, and it sounded like a snapping bone.

As the trail bent east, following the ridge, we passed through a trough of stone, walls of granite rising up to shoulder-height on either side, cool gray painted over with thin arteries of bryophyte. Bright red flowering bromeliads lit the way like luminaries. We passed through one by one, touched each other lightly to confirm that each of us had made it out alive, and continued on. There we found a fork in the trail, ragged yellow flagging tape hanging from a slanted tree blowing in the wind. To the left, the trail cut down in a steep drop along the Atlantic slope of the ridgeline; right, the path led up into green moss and mist. Tapir tracks scuffed the bank in the soft mud; we were not alone in the elfin forest after all. We rested for a few minutes at the crossroads, then followed the dark trail down along the shoulder of the ridge.

We passed a fallen tree that I recognized from Marty Crump's old photos of the breeding grounds, and Pri and I stopped to peer into its dark hollows. In tangled roots where the soil had been worn away, green moss dangled over a silent pool. I felt the stirrings of a grief for something I had never known, and I looked up to see if I might recognize the same emotion in Karen or Eladio, but they were out of sight somewhere ahead of us. There was only Kyle with his camera, kneeling low above the wet earth, filming another empty pool.

That morning, before Pri had woken up, he had asked me candidly what the purpose of this journey was: Was there something we were hoping to achieve, or learn, or find among the tangles of the past? I wasn't certain how to answer him. Part of it was only curiosity—I had longed to see these mystic forests for myself, ever since I had touched their images in faded photographs from thirty years ago. Part of it was heredity—I wanted to walk the same trails that our parents had, with my new family beside me. But there was another thought that I hadn't had the heart to speak aloud: I was wondering if there was anything left of them up here, among the mists and clouds and buried history—a revenant or a restless spirit, last remnant of the golden toads.

Pri put a comforting hand on my shoulder, and we left the little pool to follow the old trail through Brillante.

We wandered through the abandoned breeding grounds, among the trees that had been pushed over by the wind like broken statues on a road into a fallen kingdom. We passed the pools of grief where death slept at the bottom, and they looked like they might lead into the hollows of the earth. Each pool was a marker: *Here, they were here, and now they are gone.* As the wild wind howled up from the Atlantic slope, I wondered if these were their monuments or their gravestones; were we walking through a cemetery or a haunted house?

If the golden toads had become ghosts, surely this was the forest that they haunted. In clutching roots, in dark hollows, I could see the empty places the forest had held for them. Now their caverns were empty, aching, and the air had grown stale and sour, forgetting what it mourned. When the memory disappeared, there was only sorrow left. Some of the pools were washed over and filled in with leaves and soil from the tide of rain going in and out, and any secrets that they still held were buried. What might the toads have left behind when they vanished?

THE GOLDEN TOAD

As we moved through the graveyard of the golden toads, I wondered if their spirits were still, or restless with unfinished business, longing to climb back to the world. Would they know me in death, if they met me, when they had never known me in life? Would they let themselves be seen by me, when I had no right or claim to their existence? Or had their ghosts grown bitter in the long years since their annihilation, lonely in their ruined kingdom on the mountain? I imagined what it would feel like if their corpses rose from the soil and overtook me, cold fingers covering my eyes in their tarnished light as they pulled me down to join them in the earth.

What curses had I brought down on myself when I started stirring up the past? What curses on my wife? And on our future?

The mist blew off a little, and the weak sunlight touched the elfin forest. I had lingered too long in the ethereal way station, world between worlds. Pri and Kyle and Karen and Eladio had gone off with the rangers up ahead, and I could see them dimly through the twisting arms of the trees beyond. I took a last look at the forest we had come through. The earth was undisturbed and the waters were still; no restless spirits had risen from their graves. Feeling a chill, I pulled the hood of my raincoat up and walked on through the wet wind and the rippling forest.

When we reached the river, we sat on the stone banks and passed around a loaf of bread as the rangers swapped old stories about the golden toads and the Reserve back in the early days. Alvaro took my walking stick, which was a piece of bamboo I had cut in Yorkin near the Panama border, and started to whittle down the sharp tip with his machete—like a father worried for his son. Pri took off one of her rubber boots to pour the water out, she and Karen asking Eladio about the orchids he had stopped to study on the long descent from the elfin forest high above. Kyle had slipped away sometime when I wasn't looking, to peer into the shadowed canopy farther down the trail, where a black-faced solitaire sang its echoing song. In this reverie between the journey in and the return, I wandered up the wet banks along the water, following the river through green fans of leaf and listening to its quiet, constant song.

It was a private and absurd idea that I was harboring, and I wouldn't have spoken it out loud if Karen or Eladio had asked me. It had something to do with an amber glimmer in the pools of shadow along the riverbank, and dark eyes looking out from hidden dens. The golden toads had been a reclusive species, even at their most abundant; was it so absurd to think they might have evaded rediscovery, if they hadn't wanted to be found? The Brillante highlands were dense and deep: Maybe nobody had looked right here, at just the right time, when the toads had reemerged to meet the rains. For a moment, I had a vision of fortune and glory: the news going out to the world that Trevor Ritland had rediscovered the golden toad.

Beside a tall, eroded bank I stopped and knelt down to look into the last pools, water lapping at my boots. But in the holy water, I saw nothing but my own reflection. The spell evaporated, slowly, like a mist. The riverbanks were empty, no hidden gold, no glory. From somewhere up the river, Kyle's black-faced solitaire was calling, sounding like a squeaky well in the thick green jungle, drawing me back down to Earth.

When I followed the little stream back to the clearing, Eladio and the two rangers had gone off to scout the way the old trail wandered through the understory, and Karen was sitting on the bank alone. On the other side of the water, between the buttress roots of an old oak tree, Pri asked her what it had been like to walk through the Brillante highlands when the golden toads had once emerged to greet the rains.

"Well, I have a very distinct recollection of finding them," Karen replied softly, squinting up into the forest we had come through. She had walked these trails when she was just a little older than I was now, while her husband had chased tropical butterflies in the regenerating pastures. "You know, the cloud forest is a dark place. And so for us, whenever we would come across a splash of color, it would kind of wake us up and excite us. And it happened with different hummingbird-pollinated plants, or a branch of bright bromeliads . . . but there's nothing like a golden toad: so bright, and moving, in these dark, wet pools."

Looking at the heavy forest closing in around her, she studied the spitting water rushing through the narrow canyon, fingers of leaf reaching down to touch the bank.

"Now it feels new and different and a little bit unfamiliar."

After a few years in Monteverde, Karen had left the mountain at the end of 1986; by the time she returned, the golden toads had already begun to disappear.

"It's jarring because it doesn't sink in the first year," she said, remembering the empty pools in the misty forest. "You think: *They're coming out later*. It doesn't even cross your mind that they're gone. You're trying to impose your view from yesterday and all the days before; it takes a while before you put the pieces together and your new way of looking at the forest aligns. So now my concept of the forest is of a forest without frogs. But that took me a long time to get used to."

She recalled nights in the cloud forest in her early years in the country, when the amber glow from her and Alan's little cabin had called in dusky moths dancing down from the black expanse of sky, when glass frogs stuck themselves to their dirty windows, all their mysteries unfurling like the mountain's scripture. But here in the little canyon below the elfin forests of Brillante, Karen's treasured memories of the golden toads were fading. How strange that something so familiar—buried so deeply in the soil of a place—could begin to drift away.

"You know, these frogs are going to live in images, and in a smaller and smaller collective memory," Karen murmured, looking off and back up into the forest from which we had descended—where, someplace, among the dark abiding secrets of the highlands, the river was born out of the mountain. "If you could turn back the hands of time . . . it really does march forward. If you're lucky, you have the memory."

And for a second, a slide show played over her eyes—the whiff of recollection handing life back, briefly, to a forgotten moment.

INTO BRILLANTE

"It's dark . . . and then there's this bright color. And it's not just one bit of brightness—it's like this little pool of lights: *orange*."

A cloud of *manataria** fluttered out from their hidden pocket on a dirt wall, moving through us as though we were already ghosts and coloring the sky a caramel brown like smoke from a clearing fire. Karen was reemerging, piecemeal, from the shadows of the past.

"What it does is: It makes you appreciate life a little bit more—when you lose something that you realize, after the fact, that you loved. And you're sorry."

Karen told us all of this—cold from the water that had seeped into her core—while her mother was dying, three thousand miles away in the warm arms of a Michigan summer.

When the rangers picked their way back along the wet clay bank, we fell again into a winding line and left the river, moving along the hillside to climb back up to the Divide. Before long, the sound of water was lost below, and we moved up steps of soil and stone, exposed on this side of the mountain to the blowing water in the air, not yet rain but airborne mist drifting up from Peñas. We crossed an old trail leading across the valley to the San Luis waterfall, another haunted place, and turned away to climb the steep path to the ridge.

There is one place in Brillante where you should stop to leave your offerings, if you have carried them—where the trail winds around the steep edge of the Atlantic slope, and the canopy falls away but for a handful of old oaks draped in the ornaments of the gods. In that place, there is a thick mist rising from the underworld like surf, pushing green leaves over, crashing into the hearth of mountainside. It feels holy—like standing at the edge of the whole known world. When we came to this place on the old forgotten trail, we paused to feel the breath of the earth. All other sounds fell away, and only images remained: dew and morning sunlight on a spider's web; water glistening in hidden nests like

---

* Manataria: A genus of butterflies.

pearls; the path before us swallowed in the dark of vines and branches, an ageless tunnel leading into the forest's heart.

This was not a haunted forest, I understood at last; this was a forest that the golden toads had protected forever, even as they disappeared. They were not here to see it anymore—not here to meet the rains like faithful friends, not here to see the white moon rise above the mist into a sea of sky—but they had wrapped this forest in their golden arms and it was their last legacy, their living monument.

I tucked these images away, hoping that I might remember them so clear forever, as we entered again the sweet forest—where the river cut a thin line through the misty highlands, like a light under a door.

As we made our way back up onto the continental ridge, the first drops of rain were just beginning to fall; weak sunlight washed over the glistening epiphytes and lit the understory in a dirty gold. Somewhere not too far away, a quetzal sang its lonely song, peering out from its deep den in search of a mate to spend its life with. I couldn't help but wonder if, one day, one of those old deities would be the last of its kind.

At the high point on the spine, I watched Eladio pause for a moment and turn around to look beyond me. Balanced there on the high line drawn between two worlds, it felt like he was looking back into a past that only he could see: to a place in which a hundred toads the color of the sun returned his gaze with dark, illegible eyes—endemic, perennial, and wise.

When Pri and I had met Jay Savage in his San Diego home to hear the story of his first hike up into Brillante, we had asked him if he thought the golden toads might still be out there somewhere in those forgotten forests. Wading into his early nineties, Savage had held his old photo slides against the light so that he might see the toads again, one by one, and shook his head and answered with a sad conviction: "My toad is extinct."

When I put the same question to Eladio on our long walk out of Brillante, it felt like casting the die at last, for better or for worse; it was a question I had wanted to ask him for three years, ever since I'd heard his name spoken at the edge of a forest full of ghosts. And on the same ridgeline where the golden toads had disappeared, Eladio smiled softly and answered that he hoped they were. We had left the singing river far below us, and we paused to catch our breath and take shelter from the cold wind at the old crossroads, returning to the place we had begun. While the wind howled and the ribbon of yellow flagging tape rippled in the wind, Eladio told his story: hiking up to an enchanted forest farther out along the Continental Divide, farther out than anyone had ever gone looking for the golden toads before. There, he'd walked upon a high pool where a remnant clan of golden toads, so far, had escaped the great calamity—like a trove of treasure long ago forgotten, and finally discovered. He told his story as he told all others: shyly, softly, not looking to impress.

"En noventa y uno," he remembered, "yo pienso que vimos por lo menos de quince o veinte machos, y varias hembras, y mucho juveniles. Era como un montón de juveniles—en el mismo sitio." In 1991, two years after the last confirmed sighting of a golden toad on the Brillante trail, Eladio had seen fifteen or twenty males, several females, and a large number of juveniles, all together. He had carried home no photographs, no evidence of his encounter, and he had never set foot in that forest again—not in twenty-seven years.

I wished in that moment that I could see above the canopy; that I could climb a tree and break at last the blanket of the clouds, and look out along the ridge, twisting south, to see if I might achieve a glimpse of the mountain of revelation that Eladio recalled. From here, so far away, would I be able to see the green peak, its history and its long-kept secret wrapped in a concealing mist? If my vision was infallible, might I even see a flash of gold among the elfin forest on that distant ridge? And then the old pools began to seem like wishing wells, and the long trail that we had walked along Brillante felt like only the beginning of another path—a path that led through time, through death, into the last remaining sanctuary of a long-lost species, brighter than the sun.

THE GOLDEN TOAD

But I knew that no matter how it called to me, that mountain lay beyond my reach, for now. We had done what we had come to do here, after all; we had walked the old historic breeding grounds of the golden toads, we had touched the water, we had met the ghost. But our trip was drawing to a close, and in a few days Kyle and Pri and I would descend back to the city, to board a plane and fly to the United States. I understood that it might be months before I found myself in these cloud forests again, and in that time the old trails might go wild. Hope would have to be enough for now; hope that maybe they were out there somewhere, waiting, hiding, holding still; hope that on a clear night they might reemerge to gather in their secret pools in distant ranges, gold reflections dancing on the water. When darkness fell over the elfin forest and a drinking moon spilled its white glow into a twisting nest of roots, their hideouts from our changing world, when there had been no sound for hours but the whisper of the eternal wind on wet monstera leaves, when musty petrichor rose out from woody litter—maybe then the earth lit up with the glow of a hundred little fires. Eladio believed that; I wanted to believe it too.

A last wail from the mournful banshee on the wind swept up from the Atlantic slope, pushing the stunted trees into submissive bows like worshippers. We left the crossroads to follow the lonely path out of Brillante, where we would walk through wood and stone and scrape the sky along the winding trail that balanced on the backbone of the earth—then on into a damp gust, bringing rain.

(*above*) Perhaps the first photograph of the golden toads was taken by John Campbell, one of the early Quaker settlers of Monteverde who worked to preserve the cloud forests. MARTHA CAMPBELL

(*below*) A male golden toad awaits the emergence of the females in the high pools on the Brillante trail. Hundreds of golden toads emerged to greet the rains each year, then retreated to subterranean hideaways at the end of their breeding season. From one year to the next, they vanished. MARTHA L. CRUMP

(*above*) A view from the mirador in Cerro Plano. Mist and fog deliver life-giving water to the cloud forests of Monteverde, making the area one of the most biodiverse forests on Earth. TREVOR RITLAND

(*below*) The Monteverde cloud forest, somewhere below Cerro Amigos. The dark interior of the cloud forest is filled with life on life; epiphytes like moss, bromeliads, and orchids cover trees and other plants, creating a dense green forest. TREVOR RITLAND

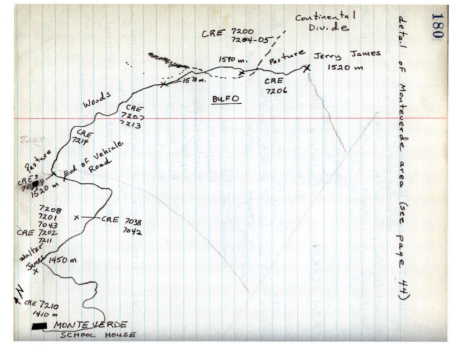

(*above*) A hand-drawn map to the elfin forest of the golden toads by Jay M. Savage, the biologist to first scientifically describe the endemic species. JAY M. SAVAGE

(*below*) The Brillante trail in the Monteverde Cloud Forest Preserve runs along a ridgeline of stunted elfin forest, twisted by the wind. The trail that leads through the historic breeding grounds of the golden toads has been closed to visitors for more than thirty years to protect the sensitive habitat. TREVOR RITLAND

(*above*) A congregation of male golden toads search for females in a temporary rain pool among the roots of a tree in the elfin forest.
MILLS TANDY

(*left*) A male and female golden toad in amplexus in a breeding pool.
MARTHA L. CRUMP

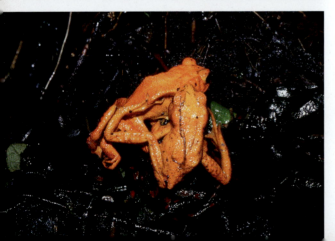

(*left*) Biologists reported intense competition between males during breeding season. "Toad 'balls' formed when up to ten males clasped each other simultaneously," Vandenberg and Jacobson wrote.
MARTHA L. CRUMP

(*right*) Trevor Ritland and Priscilla Palavicini on the Atlantic slope near Monteverde.
TREVOR RITLAND

(*right*) Kyle Ritland somewhere above the town of Cedral on a search for the golden toad.
TREVOR RITLAND

(*below*) Luis Solano, Eladio Cruz, and Mark Wainwright rest during a return to the site of the golden toad's last appearance, thirty years after Eladio's last sighting.
TREVOR RITLAND

A pair of harlequin frogs at the Monteverde Cloud Forest Preserve, before the local amphibian declines. The Monteverde harlequin frogs at the Preserve and Rio Lagarto were previously classified as *Atelopus varius*, but they are now believed to be a different species entirely. MARTHA L. CRUMP

The emerald glass frog (*Espadarana prosoblepon*) declined locally around Monteverde, but small populations are beginning to return. The authors encountered this individual during the 2021 search for the golden toad.
TREVOR RITLAND

The green-eyed frog (*Lithobates vibicarius*) was widespread in Monteverde and the surrounding forests until the amphibian declines of the late 1980s, when it vanished for more than a decade. In 2002, Mark Wainwright rediscovered the species in the Monteverde Cloud Forest Preserve, and populations are continuing to recover. The authors encountered this individual during the 2021 search for the golden toad. TREVOR RITLAND

The biologist Juan Abarca rediscovered Holdridge's toad (*Bufo holdridgei*) as a student on a field trip with the University of Costa Rica, after the species was previously classified as extinct. It is the closest living relative to the golden toad. JUAN G. ABARCA

(*above*) The narrow-lined tree frog (*Isthmohyla angustilineata*) is a critically endangered species with less than 250 mature individuals believed to exist in the wild. A team from the Monteverde Cloud Forest Preserve discovered a small population several years ago, and three individuals were recently discovered in Monteverde's Children's Eternal Rainforest. "It's not orange, and it's not famous," says Mark Wainwright, "but if we found three, and they've been gone for thirty years, then there's a sustained population—one that is resilient to whatever calamity we may attribute the decline to." TREVOR RITLAND

(*right*) Donald Varela Soto discovered the Tapir Valley tree frog (*Tlalocohyla celeste*)—a species new to science—after converting a former cattle pasture into a nature reserve. Soto is now working to protect the new—and critically endangered—species with support from the local community.
MARCO MOLINA

This is the last photograph of a golden toad ever taken. Frank Hensley took the photo in the Brillante breeding area on May 15, 1989. This male toad was the last ever seen in Brillante, and is widely reported to be the last golden toad ever documented before their disappearance. FRANK R. HENSLEY

The first golden toad collected by Jay Savage, now housed in the herpetology collections of the Natural History Museum of Los Angeles County.
NATURAL HISTORY MUSEUM OF LOS ANGELES COUNTY (LACM)

Some of the 142 male golden toads collected on Savage and Scott's Costa Rica Expedition, 1964.
NATURAL HISTORY MUSEUM OF LOS ANGELES COUNTY (LACM)

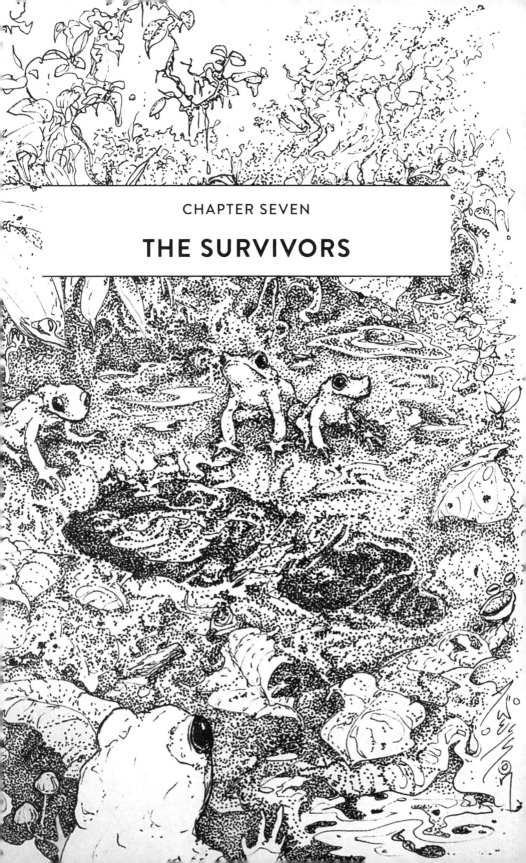

CHAPTER SEVEN
THE SURVIVORS

Monteverde, Costa Rica

There are places where things have come back from the dead. For instance, there is a little-used trail that runs from Monteverde to the Arenal volcano along a ridge, a two-day hike through the forests of the Atlantic slope. In the rainy season of 2002, Mark Wainwright was hiking along those steep trails to the foot of the volcano when he saw a ghost.

A British naturalist who had visited Monteverde in the 1990s and never left, Wainwright had become a local educator and a guidebook author. He had spent several years conducting surveys for the lost frogs with Alan Pounds, but because he had arrived in Monteverde after the amphibian crash, he had never had a chance to see the most famous species for himself. It was a bit like arriving late to a celebrated show, just in time to see the curtain close.

Moving at a good pace down the steep trails, Wainwright realized that he had gotten ahead of his hiking partner, so he stopped for a little while in a wet and puddly area a little off the path.

"So I thought I'd sit on a log and wait," Wainwright remembers, "and out jumped this big frog."

The frog was a bronze-brown color, with a white lip stripe and a dark band running down its body. It had bright green irises, almost iridescent.

"And I thought, 'Goodness me, that's a green-eyed frog. I thought they were supposed to be gone—how cool!'"

THE GOLDEN TOAD

Along with the golden toads on the ridge, the harlequin frogs at Río Lagarto, and more than twenty other species from the cloud forests around Monteverde, the once-abundant green-eyed frogs (*Lithobates vibicarius*) had disappeared in the late 1980s and the early '90s; it had been more than a decade since any survivors had been reported.

Heartened by the auspicious omen, Wainwright continued on until he came across a large pool in the forest interior, where he discovered a mass of sizable tadpoles. "I mean, there were a gazillion of them," he remembers. The only group of frogs with tadpoles that large, Wainwright knew, were the Ranidae—the true frogs. And at that elevation, there was only one likely option. He collected several tadpoles, and, when he made it back to Monteverde, he hiked up to the foot of the Reserve to deliver them to Alan Pounds. In a few weeks, the tadpoles had grown into juvenile green-eyed frogs, and Pounds sent word to Wainwright that—in the same forests that the golden toads had long abandoned—a lost species had come back from the dead.

In the following year, an equally unexpected encounter occurred in southeastern Costa Rica, in the mountains of the Quepos Biological Corridor. It was July 2003, and a pair of young biology students named Justin Yeager and Mark Pepper had been spending the last few months searching for a green-and-yellow color morph of the poison dart frog (*Dendrobates granulifera*).

"It was our white-whale trip," says Yeager. "We'd been all up and down the mountains looking for them, up and down every road we could find. We'd destroyed two or three gears on our car."

In the Fila Chonta Mountains, the owner of the local Rainmaker Conservation Park had lured them to the forest reserve with tales of the frog they sought; but it was word of another frog entirely that persuaded Yeager to take the owner up on his invitation.

"He said, 'We have this other one called the variable harlequin frog,'" Yeager remembers. "And I thought, 'That's the common name for two species, neither of which should be here.'"

THE SURVIVORS

Yeager was ready to dismiss the owner's claims as total fiction, but Pepper convinced him that the possibility was worth investigating. As they traveled up to the reserve, Yeager's doubts only grew heavier.

There's no way that one crossed from Central-Southern Colombia, through the Darien Gap, through Panama, and is in Pacific Costa Rica, he thought to himself. *There's no way. And the other one is even less likely, because it's extinct.*

While more than a hundred populations of *Atelopus varius* had once been known to Costa Rica, it had disappeared from its highland habitats* at the end of the 1980s, and by 1996 it was considered to have vanished from the country altogether. Surely, Yeager thought, it wasn't hiding in a terrarium in a private reserve.

But when the owner revealed the toad, Yeager couldn't believe what he was seeing. He dashed off to call Federico Bolaños, his adviser at the University of Costa Rica, and when Bolaños answered, Yeager told him: "I'm looking at *Atelopus varius*."

Bolaños responded: "Bring me a toad!"

Before long, Yeager had a tissue sample preserved in a salt-shaker full of alcohol, and within a few weeks Bolaños's DNA analysis had confirmed what Yeager had been unable to believe before he'd seen it for himself. The variable harlequin frog had returned.

Five years later, in May 2008, a student named Juan Abarca was on a field trip with his University of Costa Rica herpetology class in the oak forests of Volcán Barva in the central mountain range north of San José, when he observed eight tiny toads in the leaf litter among the grass. Only two species of bufonids had ever been reported at that location: the notoriously secretive Chompipe spiny toadlet (*Incilius chompipe*), and Holdridge's toad (*Bufo holdridgei*). The latter had disappeared from

---

* It is worth noting that recent hypotheses have proposed that the Río Lagarto harlequin toads were in fact an entirely different species from the variable harlequins (*Atelopus varius*), not formally described by science. If this is true, it means that the Monteverde harlequin toads disappeared before biologists could even give them a name.

the area in the 1980s, and after twenty-five years of fruitless searches, the International Union for the Conservation of Nature (IUCN) had finally classified Holdridge's toad as Extinct in 2007.

They photographed the juvenile toads, but the photo quality proved too poor to identify the species. "For six months, we were dreaming about it," Juan Abarca recalls. "Could it be an extinct species? We had dreams and nightmares about the idea."

So, in April and May of the next year, Abarca returned with a small team to look for the mystery toads again—searching the leaf litter, peering under logs and rocks and mossy banks, wading through the wild grasses and deep ditches on the slopes of the volcano. Thirty years ago, the area had been mostly pastures and open fields, but now regenerating forest was recolonizing the landscape; tall oak trees rose above dense undergrowth where shallow puddles and temporary pools collected from the rains, curtained by ferns and bamboo. At a little pond along the roadside, they collected several tadpoles to take back to the university, hoping to identify them as they matured.

Finally, at the end of April, the team found one adult male toad (documented as UCR 20671) hidden in the leaf litter, just five meters from the place where they had found the eight juveniles the year before. They brought the specimen to the Zoology Museum at the University of Costa Rica and presented it to Gerardo Chaves, the museum's curator.

"He said it was a ghost," Abarca remembers. "Holdridge's toad."

Over the next few days, they found more: juveniles and adults on an abandoned trail covered with grass, a lonely male two centimeters beneath the earth. The extinct Holdridge's toad had come back from the dead.

The resurrection of the toad came fifty-seven years after its first description, when the herpetologist Ed Taylor had encountered a single specimen at the Finca Georgina on the western slope of Volcán Barva in 1952. Subsequent populations were later identified in the montane rainforests on the southern and eastern slopes, and biologists hypothesized that the toads were probably fossorial—living underground for long periods, beneath mossy banks and in drainage ditches beside the roadside paths, and emerging in large numbers during rainy spells. In the 1970s, 2,765 males had been recorded at the highland breeding pools over an eight-day period. The males were observed to wrestle with one another to achieve amplexus with the females, sometimes creating roving balls of clambering toads. Some even reported that the toads took on a dusky orange appearance during breeding season.

If these characteristics strike you as familiar—a secret life beneath the earth, an auspicious revival to greet the rains, explosive breeding, competition, a golden glow, a vanishing—they do so with good reason: The newly rediscovered species was the closest living relative to the golden toad.

With the mystery species finally identified, Juan Abarca and his team once more climbed the slopes of the volcano to return the tadpoles they'd collected to the roadside pond. There, they watched the future generations swim off into dark waters through the mists. There was work ahead of them yet, to ensure that the reappearance of this long-lost species was not a wasted trick—but for the moment it was enough to know that they were out there in the young oak forests wrapped in mist, and that they might have other secrets still to tell.

These were not the first stories of reemergence after death. In the Indigenous communities of Yorkin and Boruca—on the Atlantic and Pacific slopes of the Talamanca Mountains, where I had traveled in my early days in Costa Rica—I had heard myths of frogs as caretakers of newborns, curers of diseases, heralds of calamity, ambassadors of the rain. I had also heard stories about resurrection and rebirth; pieces of the older wisdom were reflected in ecology, and science floated next to myth in dark waters. We had all seen amphibians work miracles—tadpoles transforming into frogs, toads appearing from beneath the earth at the ends of dry years. Was metamorphosis so dissimilar from rebirth? Was reemergence all that different from resurrection?

Now, those old stories made me wonder about the golden toads. If they'd had any miracles left in their old stores when extinction had come calling—any magic tricks—had they been enough to conquer death?

Washington, D.C.

I was looking at a golden frog on the streets of Chinatown in Washington, D.C. It was Jin-Chan, the luck-frog, and it was gazing at me through the steamy window of a little restaurant on 7th Street, coin in its mouth, promising good news. It was a fortuitous encounter. Good news was what I had been looking for—ever since the buried spirits had started stirring once again.

Eight months earlier, Pri and I had left the cloud forest and flown back to our home in Arizona, where we had gotten married in a little chapel north of Flagstaff and put the hand-drawn maps and memories away. I had gone back to the classes I was taking at the university, showing off my documentary about the golden toads on an old projector, and my thesis committee had considered the university's money well-spent. Pri had watched the snow fall through the windows while she waited for the government to approve her work permit. In October, we drove to northern Utah and watched Marty Crump draw out old photo slides and field notebooks from her last hike into the mystic forest: "GOLDEN TOAD!" scrawled in faded pencil on a rain-worn sheet, her last encounter with the lost species. We were filling in the missing details, closing all the open questions of our journey through the past.

But as hard as we had tried to bring the story to a close, there was still one thorn sticking in my skin: Eladio's encounter. Ever since he had told his story on the trail down from Brillante, I had been reading about resurrection. I'd spent sleepless nights in the glow of my laptop, piecing together the disparate accounts of lost frogs coming back to life. And by accident, I had located the origin of the talisman that had reawakened my obsession with the golden toad: the little handbill, "Most Wanted," that I had found at the biological station in Monteverde back in 2015; its architect had been a conservation biologist named Robin Moore. One night, when dreams of golden toads had haunted me, I'd sent an email to the man I'd never met to tell him that I'd heard a rumor of a later sighting, two years after Frank Hensley had photographed the last toad on Brillante in 1989. Fourteen hours later, Moore replied: "This sounds fascinating. Just the idea that the Golden Toad could be clinging on gives me chills." So, in February 2019, while Pri flew back to Costa Rica to lead a field course through the Caribbean forests, I landed in Washington, D.C., and made my way through the concrete jungle to hear Robin Moore's stories of "Lazarus Species"*—frogs that, seemingly, had come back from the dead.

I met Moore at the Global Wildlife Conservation offices that looked down over Chinatown, and we found a quiet place to talk. As he told me stories about collecting frogs and toads in the Scottish Highlands of his childhood, he revealed a familiar attraction to the unknown.

"Growing up," he told me, "in Britain, you would hear these stories of 'the Beast of Bodmin Moor.' These cats, big cats." In the 1980s, locals had reported more than fifty sightings of a five-foot-long black panther with white or yellow eyes on the rugged Bodmin Moor in English Cornwall. "And I loved the idea that there were big cats in Britain, because it just brings something wilder to the place. I mean, I grew up in Scotland, so we've got Nessie. And I love that people still go out there."

Not long after Moore had finished his Ph.D., in the aftermath of the cataclysmic declines in the 1980s and '90s, the results of a global amphibian assessment were published: A third of amphibians were threatened with extinction. As a child, he had collected frogs and newts, watching their transformations in tanks in his bedroom; he had never considered that they might disappear before his eyes.

---

* The "Lazarus" nomenclature is derived from the biblical story of Lazarus of Bethany, who was said to have been raised from the dead by miracle four days after his death.

"Growing up, you don't think about the possibility that they could be gone," he said. "You study the dinosaurs, which have gone—and here we have these living, breathing animals that are at risk of going the same way as the dinosaurs. So, for me, that was a real wake-up call. That was alarming."

As a biologist, Moore understood the very real threats that were facing the planet's vulnerable amphibians; on the other hand, his fascination with the local folklore from his childhood granted him a different perspective. To believe in something like the Beast of Bodmin Moor, or Nessie—or to even *want* to believe—required something that seemed to be in short supply in the wake of the amphibian extinctions: hope.

"And the search for lost species is built on that idea of hope," Moore told me. "That all is not lost."

Moore's own journey on the trail of resurrection began in 2010, when he was working as an Amphibian Conservation Officer with the nonprofit Conservation International. The accounts of rediscoveries in Costa Rica had been adding up—the green-eyed frog, the variable harlequin frog, Holdridge's toad—and the news had made its way to Moore while he was trying to drum up awareness about the amphibian extinction crisis. It had proven to be difficult work.

"It was hard to be that depressing guy at a party," he remembered, "when you're just telling people that things are going to shit." But in the shadow of the ongoing amphibian declines, Moore began to see reasons for hope. In northeastern Queensland, Australia, the young biologist Robert Puschendorf had recently rediscovered the armored mist frog (*Litoria lorica*), last seen seventeen years earlier in an entirely different habitat. And in the Southern Tablelands of New South Wales, a fisheries conservation officer had reported the first yellow-spotted bell frog (*Litoria castanea*) to be seen in thirty years.

Australia . . .

Costa Rica . . .

THE SURVIVORS

By all accounts, understanding of the amphibian crisis had grown out of Australia with the disappearance of the southern gastric-brooding frog, and the wildfire had burned its brightest with the extinction of the golden toad in Costa Rica. Now, in the last seven years, lost species from both countries were reemerging. Perhaps these amphibians were more resilient than previously believed.

In February 2010, Moore and his colleagues were wrestling with the challenge of inspiring public engagement with the amphibian crisis when so much of the news was daunting and dispiriting. With the recent accounts of rediscovery in mind, they couldn't help but wonder: What if there were others that were still out there? Not only did these reemerging frogs offer an opportunity to engage the public in an exciting and positive way, but they also offered clues to the mysteries that had haunted biologists for decades.

"If there are survivors," Moore wondered, "are they genetically resistant? Are they living in an area that's helping them survive? We have all these questions, unanswered, and if we can actually find those that survived, maybe they could hold a lot of answers."

With lost frogs on their minds, Moore and his colleagues put out a call to amphibian experts: Which species from their regions had gone missing, and which lost species would be the most valuable to find? Nominations from biologists around the world poured in, and it wasn't long before the group had compiled a list of more than one hundred lost frogs. The causes behind their disappearances were various—environmental degradation, remote and hostile habitats, the chytrid fungus, and climate change—and the dates of their last sightings were scattered: Some species had been documented as recently as 1995, while others had not been seen for 140 years. When Moore bumped into Rob McNeil—Conservation International's media manager—wandering the halls one afternoon, he pitched the idea for a search for the world's "Most Wanted" lost amphibians. McNeil replied, "I fucking love it."

Over the next few weeks, they coordinated with local scientists to organize more than 120 researchers on 33 search teams, with 40 lost frogs targeted across 21 countries. At the same time, Moore was working with Conservation International's communications team to whittle their initial catalog into a top ten list of the most wanted.

"The idea behind the top ten was to hook the most visual creatures of them all," Moore would later write. "We needed our charismatic frogs to be our flagships for the campaign, an entry point for people to learn more about the plight of the less colourful or less charismatic creatures."

The gastric-brooding frog found a place in the top ten, acknowledging the lost Australian amphibians. The tropical harlequin frogs were also represented in the Rio Pescado stubfoot toad from Ecuador (last seen in 1995) and the scarlet harlequin frog from Venezuela (known only to a single stream site). Then there was the Hula painted frog from Israel, last seen in 1955; the African painted frog from Rwanda, which nobody had ventured to look for in over fifty years; and the Mesopotamia beaked toad from Colombia, representing frogs lost to armed conflict. Only two individuals of the Sambas stream toad from Borneo had ever been documented, in the 1920s, and no photographs had ever been taken. Two species of salamanders had also found a place: the black-and-yellow Jackson's climbing salamander, which the biologist David Wake had described as "the most beautiful of salamanders," and the Turkestanian salamander—the oldest lost species on the list, whose range was unknown, represented only by a black-and-white drawing and last seen in 1909. And at the top of the list, in a place of honor, the number one most wanted: Costa Rica's golden toad.

"I think the golden toad is kind of a poster child for extinction," Moore reflected later. "It's this bright neon orange toad—you know, it looks like it's had a bucket of paint poured over it. So it's striking; it's beautiful. And I think its story is just poignant."

With their list narrowed down to ten species, and with teams of biologists ready to begin the search, Conservation International announced the campaign in the summer of 2010 with the release of a handbill reading "Wanted Alive."

In spite of his personal enthusiasm for the expeditions, Moore spent the early days of the Search for Lost Frogs tracking the progress of the international teams from his computer screen; the Australian team was using mapping technologies to identify remote habitat that might enable the northern gastric-brooding frogs to survive alongside chytrid; in Borneo, local development had enabled the team to access previously un-surveyed mountain summits. But every step forward came with another step back, and it was clear that none of the expeditions would have an easy path ahead of them. In the first few weeks, the local teams reported tropical storms, landslides, drug traffickers, death squads, and saltwater crocodiles. When Moore joined the search in person, his expedition moved through forests controlled by the Revolutionary Armed Forces of Colombia and checkpoints manned by teenage soldiers bearing AK-47s. Despite discovering two undescribed frog species, they never found the Mesopotamian beaked toad last seen in 1914. "All of Colombia's lost frogs," Moore would write, "twenty-two in all—remain lost."

But in early September—a month into the campaign—the fortunes of the search began to turn. Biologists in Africa's Ivory Coast and the Democratic Republic of Congo sent word of two lost frogs that they had rediscovered. Soon afterward, a young biologist in Mexico wrote to report his rediscovery of the cave splayfoot salamander—not seen since its first description in 1941—in a limestone cave outside Durango; his search had also turned up a second salamander that the IUCN had classified as "possibly extinct."

"Then each rediscovery got attention," Moore told me. "We had boots in the mud, people going out looking, and it really just snowballed." In October 2010, Moore's next expedition brought a bounty. "I went with a team to Haiti," he remembered, "and we found six frogs that hadn't been seen in twenty years."

THE GOLDEN TOAD

Later that same month, the news came in from southern Ecuador that a family of local farmers had led the search team to a streamside sighting of a black-and-yellow frog: It was the Rio Pescado stubfoot toad—the first of the top ten most wanted to be rediscovered.

When Conservation International's "Search for Lost Frogs" campaign drew to a close, international teams of biologists and local experts had reported the rediscovery of fifteen lost frogs—all previously believed to be extinct.

"I mean, out of those thirty-three expeditions, 10 percent of them were successful," Moore reflected later. "So the hit rate wasn't great. But, you know, there's a reason they're lost. So that was really the spawning of the search for lost species—and it was a six-month campaign, but the idea just lived on."

In July 2011, five months after the end of the campaign, the Borneo team emailed Moore to announce that they had rediscovered another top ten species: the sambas stream toad, which they would rechristen the "Borneo rainbow toad" in honor of its multicolored, otherworldly visage. It was the first of its kind to be seen in eighty-seven years, and the team had discovered it on a high tree trunk, a few hundred meters above the elevation that it had previously inhabited; it is likely that the changing climate had driven the persistent species higher up the mountain. And in November of that year, Moore received more surprising news of another top ten species: The Hula painted frog from Israel, not seen for fifty-five years, had been rediscovered by a warden at the nature reserve.

When I met Robin Moore in February 2019, it was eight years after the original "Search for Lost Frogs" campaign had ended. But since 2017, Moore had been working with Global Wildlife Conservation to carry the torch as the "Search for Lost Species," refining a database of twelve hundred nominations into a list of the top twenty-five missing mammals, birds, fish, insects, plants, reptiles, and amphibians. Within a year, the first of the top twenty-five—the Jackson's climbing salamander in Guatemala, known as "the golden wonder"—had been recorded for the first time in forty-one years.

A month before I spoke with Moore, Global Wildlife Conservation had announced that a local team from the Museo de Historia Natural Alcide d'Orbigny in Bolivia had finally located a mate for "Romeo"—a male Sehuencas water frog that had found fame on social media as "the world's loneliest frog." After more than ten years with only a single known individual in captivity, there was hope for the disappearing species once again.

"And last week," Moore continued, "we have what appears to be the Fernandina Giant Tortoise—lost for one hundred and thirteen years—found in the Galapagos. I mean, if you can find a missing giant tortoise after a hundred and thirteen years. . . ." He trailed off, envisioning the unspoken possibilities. "I'm very reluctant to write anything off."

Before the year was out, news would come in of two more significant rediscoveries: The first was the Mindo harlequin toad in the cloud forests above Quito, Ecuador, not seen since 1989. When the frogs were tested for the chytrid fungus, they showed no signs of infection—they had either escaped the virus, or developed an immunity. The second was the starry night harlequin toad of Colombia's Sierra Nevada de Santa Marta, documented for the first time since 1991. While the toad had been lost to science for close to thirty years, it was never lost to the Indigenous Arhuaco community, where the local people venerated the species as a guardian of water and a symbol of fertility. Four years of dialogue between the Colombian NGO Fundación Atelopus and the community's spiritual leaders had resulted in a blessing for the team of biologists to travel to the remote habitat to see the toad, without taking photos; a test to determine if their intentions were pure, and if they could resist temptation. The local biologists reported a population of thirty "lost" starry night toads, alive and well.

"Local knowledge is key," Moore had told me in the months before the news of the starry night toad would emerge. "I think it's arrogance to say that we go around the world rediscovering frogs, because that wasn't the ethos behind the Search for Lost Frogs. The search teams that we were supporting all relied on local knowledge and, where possible, were local teams. And often you go into an area and look for a lost species and the local people are like, 'It's not lost.'" He told his own story of looking for a long-lost harlequin frog in the páramo* of Ecuador, receiving directions from local people on the side of the road, them pointing up into the forest, saying: "Yeah, a little black frog, a couple of hours that way." Lost to science, it would seem, was not the same as lost.

"So I think accepting that there's a lot out there that we still don't know is really important," Moore said. "And Lazarus Species hold that possibility of something kind of magical and mythical becoming real."

In a lot of ways, the possibility was more powerful than the certainty. He wanted to believe in Nessie, of the deep lochs of his childhood, but he didn't want to go and look for her. "I hate the idea of doing these sonar sweeps," he said, "because I almost don't want to know that it's not there. So I think Lazarus Species hold a special place for us, and for me especially; it's just this allure of: What we don't know is more interesting to me than what we do know. And I think as I've studied more, I just realize how much we don't know."

The gray snow was blowing in on Chinatown outside, and the morning was growing late. I thanked Robin Moore for his time and stories, and we ate a quick lunch together in the busy restaurant on the street before he disappeared into the offices again. I spent the afternoon wandering the Smithsonian museums, looking at the taxidermied animals preserved in time, all their mysteries revealed. The next morning, I flew home to Arizona, but I would keep tabs on the Search for Lost Species from a distance, and when Global Wildlife Conservation partnered with Leonardo DiCaprio in 2021 to become Re:wild, they released a

---

* Páramo: A high-altitude grassland ecosystem.

list of ten "Lost Legends." These were the species that lived in a collective imagination, that had been lost for so long that their rediscoveries were classified even by the people looking for them as "long shots at best." The first three species on the list were the Tasmanian tiger, the ivory-billed woodpecker, and the golden toad.

I wasn't so sure about the long shot. With Robin Moore's stories added to the growing list of resurrections, I was beginning to believe that disappearance did not always mean extinction, and I often found my thoughts returning to a group of golden toads seen at a far pool on a lost mountain years after their retreat. But reemergence was one thing; survival was another. If there was hope for the lost frogs of Monteverde, they would have to do more than reappear; they would have to learn to survive again in a deadly world, among new perils that had driven their relatives into annihilation. For that, I knew, the secrets lay in the stories of the frogs that had returned, and the local people working to protect them.

Monteverde, Costa Rica

An electrical storm came in with the heavy rains, and the tents flooded, and the water rose in the pools at Chutas. Two nights ago, Mark Wainwright and the three herpetologists from the Ranario in Monteverde had been warm and dry at their midway camp, where they had slept in the old cabin that had been closed and off-limits for more than a year. That night, the abandoned cabin had seemed eerie and unsettling; now, as the team fled to higher ground to avoid the pooling water, their former lodgings didn't seem so bad. But even the storm with its bright lights and booming thunder couldn't darken their spirits. Earlier that day, they'd found several hundred green-eyed frogs in and around the pond that Wainwright had reported close to six years earlier, deep inside the Bosque Eterno de los Niños (BEN), adjacent to the Cloud Forest Preserve. Even as the storm crashed against their tents, they could hear the frogs calling in the night, singing with the joyful voices of the resurrected. For the faithful who had held out hope for the lost frogs of Monteverde, the rediscovery was only the beginning of the work to come. Ensuring the long-term survival of the imperiled species would require local and international collaboration, a captive breeding program, and investment in the ecosystem and community that had enabled the frogs to endure in the shadow of other extinctions.

"After finding the green-eyed frog, I mentioned it to the guards of the reserve like it was a big deal," Wainwright remembers, "and their response was: 'Yeah. That frog's always up there.' But they didn't know that it was called the green-eyed frog, and that it was supposed to not be here. Why would they?"

When Wainwright hiked back to the site in September 2008, the group included park guards and other staff from the local forest reserves who had demonstrated an interest in the project. In addition to taking more water samples, testing pH levels, and replacing the data loggers, the team also participated in field workshops in frog identification and sampling protocols. Even after a long hike and a rainy evening spent digging drainage ditches around the camp, the forest guards were so eager to put their new knowledge to the test that they set out on a midnight excursion to look for frogs.

The next month, Wainwright helped coordinate a two-day amphibian workshop at the BEN's San Gerardo Field Station for local rangers. Conducted entirely in Spanish, the workshop covered identification techniques, natural history, taxonomy, and a history of frog declines.

"There's no one who comes close to spending as much time in all these areas as the guards and the maintenance crews," says Wainwright. If their conservation plan was going to succeed, the resources and support from international collaborators would not be enough. The future of the green-eyed frog would depend on the local people. That future seemed to be in good hands when, several months after the San Gerardo workshop, the forest guards were still stealing away on their lunch breaks and evening shifts to climb into the highlands and "go frogging."

Two years after Juan Abarca rediscovered Holdridge's toad, conservation efforts for that species were underway as well. The local team began monitoring spawning pools, population dynamics, and chytrid infections in the toads at Alto El Roble, creating artificial pools to encourage reproduction. Abarca conducted workshops in local elementary schools, in small towns where the parents remembered the toad from decades ago but had never known its name. But even with these efforts, the future of the toad was still uncertain. Abarca had trouble finding students who wanted to carry on his work with a species that was so difficult to locate; after all, the toad had returned from presumed extinction "por pura chiripa"—only by luck. The area in which the toads had been found was a popular (but illegal) tourist destination, with thousands of visitors each year roaming the trails on foot, bikes, and ATVs, and the forest was so fragile that even biological study proved perilous. Abarca worried that he was destroying the habitat bit by bit with every hour he spent searching for the toads. Even after the reemergence, the toads had only been found at two sites, two kilometers apart: a small pond next to Rio las Vueltas, and another on the path to Cerro Dantas. Those ponds were growing drier, and entire generations were being lost. The habitat would have to be protected, and biologists would have to learn more about the mysterious species, or Holdridge's toad would be doomed to disappear again.

Resurgence, you see, does not guarantee survival.

Surveys beginning in 2005—two years after Justin Yeager had encountered the variable harlequin frogs in the forests of Fila Chonta—yielded no additional populations of the lost species in the area. At most, no more than thirty-one individual toads were believed to survive at the Fila Chonta site. Fortunately, a second population was discovered in the Las Tablas Protected Zone near the Panama border in 2008. One individual was later reported near Buenos Aires (not far from the Boruca community), and other populations were documented in western Panama. As recently as 2015, the herpetologists César Barrio-Amorós and Juan Abarca reported the discovery of another group of survivors in the general area of Uvita de Osa on Costa Rica's Pacific coast. They did not provide the exact location for fear of "dangers from illicit collecting or the potential introduction of disease." After hearing that "yellow and black frogs" had been sighted along a stream in a private reserve, Barrio-Amorós accompanied a guard to the primary forest habitat, where he observed ten adult male variable harlequin frogs: nine alive, one dead. The deceased individual later tested positive for *Bd*, and, despite intensive fieldwork, Barrio-Amorós has not yet encountered a second population in the area. "Of the 169 localities recorded in the University of Costa Rica database," he wrote in 2021, "only seven are known to survive." A new population was discovered later that year, but only a handful of those eight populations are likely to be viable long term; the reports from Fila Chonta and Buenos Aires have gone silent, and deforestation (for palm oil and pineapple and livestock), pollution, the pet trade, disease, and climate change cast a shadow on the others. Despite their reemergence, the fate of the species hangs by a thread, and at the time of this writing, no harlequin frog has been seen in Monteverde since they abandoned the Lagarto River thirty-five years ago.

But the cloud forests of Monteverde mourning for the spectacled frogs that once basked in the sun by the river might be comforted by this news: Today, there are green-eyed frogs in a captive breeding program at the Chester Zoo in England—like babies tucked in baskets and sent downstream to wait for better times—and the wild population between the Continental Divide and the volcano is alive and well. Mark Wainwright reports that as you near the pool, you begin to see them on the trail, and when the rains call them out from the wet forests, they gather by the hundreds in the standing water.

THE SURVIVORS

• • •

When I talked with Robin Moore in 2019—nine years after the Search for Lost Frogs began—one of the most compelling mysteries remained. The poster child of extinction, the number one most wanted, had yet to be reckoned with: the golden toad.

"We didn't specifically organize an expedition for the golden toad," Moore had told me. "But I was in close touch with people on the ground in Costa Rica who were best placed to look for it." One of those people had been Alan Pounds—who, after twenty years, was still searching. "I said something like: 'So you think there's a glimmer of hope that it's still there?'" Moore recalled. "And I think he replied: 'More than a glimmer.'"

Alan Pounds had said something similar to Pri and me when we had talked with him in the casona at the Reserve a few days before we'd climbed into Brillante. Every time he hiked into the highlands, or when he rounded the corner at the end of the Nuboso trail just below the Continental Divide, he looked for them—old friends, finally come home. If the others could reemerge—the green-eyed frogs, the harlequins, Holdridge's toad—why not them?

"It's possible that they could be in some of the less accessible areas out there," he told us. "Anything is possible. It's hard to know when to declare something extinct. Because as soon as you do, maybe that's when it reappears."

And Alan Pounds was not the only one who still had hope for Monteverde's ghost. Although the original Search for Lost Frogs had long since drawn to a close, Robin Moore still talked about the golden toads like they were unfinished business.

"You know, depending on who you talk to, you get a very different perception," Moore had said. "Anyone I meet from Costa Rica, I ask: 'Do you think the golden toad still exists?' And it's always fascinating what the answer is."

THE GOLDEN TOAD

Karen Masters, for instance, left Monteverde in 1988, and when she returned a few years later, the golden toads had disappeared. But in 2001, ten years after the decline, she followed the cordillera all the way out to Cerro Bekom—an elfin forest on the ridge at the same elevation as Brillante. "And we went around May 15," she remembered as we walked on the Brillante trail together, "figuring, you know—maybe—we'd find golden toads."

Twenty-nine years after his two-week study, Mills Tandy was still waiting for news of the vanished species. "There's a lot of montane elfin forest that nobody goes to," he told me when I asked him if he thought the lost species might return. After all, other frogs that disappeared around the same time as the golden toads, he reasoned, had been found. "It's peculiar that the golden toads have not," he said.

When I met Marty Crump in her Utah home in October 2018, I had asked her the same question. "I have to admit that until about a decade ago," she'd said, "people would ask me: Do you think the toads might still exist? I always said 'yes.' If in fact it was the chytrid, or at least in part the chytrid, it's conceivable, maybe, that there are some areas where the toads were not affected. And maybe they've been able to survive in their little tiny localized area. When people ask me now if I think the golden toad is extinct, I say 'probably.' But I don't think I've ever said 'yes.'"

"It's the only thing I ever saw that nobody else saw after," Frank Hensley had told me, recalling the last golden toad ever documented on Brillante. He still had the photo he'd taken of the little endling, never published, last of its kind. "I'd like to unbreak my heart. I'd love for the golden toad to be something that one day I could take my son to see. Few things in this life would make me happier than that." When I'd asked him if he had hope that the golden toads could still survive, Hensley had replied: "I have a thread of hope. I hope for hope."

Others are more optimistic. Mark Wainwright, who rediscovered the green-eyed frog in the same forests that the golden toad had once called home, knew as well as anyone how easy it was for lost frogs to remain hidden. "In those remote areas, how many people ever go there?" he wondered. "I don't think I would ever rule out the possibility of there being any of those species in this vast and mostly unexplored forest."

When I ran into Giovanni Bello one morning at the Reserve, where he was working as a tourist guide, I asked him what the return of the golden toad would mean for Monteverde; he'd replied: "Protección absoluta." Total protection for the cloud forests sheltering not only golden toads, but thousands of other species too. "Mucho mejor for everybody," he had said. "It's endemic. It's our toad." In the early 1980s, he had spent long days in the highlands searching for the pools where the toads had once emerged to greet the rains, faith and loyalty eternally rewarded, until the calamity had fallen over them and they could be found no more. When we stood together at the foot of the Reserve, Giovanni looked up at the green mountains lost among the mist and added: "It is my hope that the golden toad is waiting for me."

"Coming upon them in their groupings around tiny temporary pools, while cutting and measuring boundary lines, cold and soaking wet in the howling wind, was a huge source of joy and wonder," George Powell told me in a text message. "I can only hope that somewhere up on those mountaintops, in their own private world, a small population is still thriving."

When Robin Moore and I had eaten lunch together in the little Chinese restaurant below his office in Washington, D.C., I'd put the question to him as well: Almost thirty years had passed since the last sighting; did he still hold out hope that the golden toads survived? He paused for a long time before he answered, thinking carefully about how much of himself he wanted to reveal. What did he think about it now, and what would he have thought about it as a ten-year-old, raising tadpoles in a few inches of water and peering through the dark for signs of a beast on Bodmin Moor? "As long as there's habitat," he'd said at last, "as long as there's little pockets where people haven't been . . . it's possible it could be hanging on, you know, somewhere. So . . . I still hold out hope it could be there."

And then, of course, there was Eladio. *En noventa y uno . . . un montón de juveniles. . . .* The smell of rain had been on the wind when we stood on the ridgeline of Brillante, having walked into and out of the past. Eladio had helped draw the boundaries of the Reserve, had helped carve out the paths that we had walked down to the river and back up into the clouds. If anybody knew these forests, it was him. Others had left the cloud forest when the golden toads disappeared, never to return; others had gone away and come back from time to time; but Eladio had stayed behind, night watchman, standing guard. The years had worn him down a little, time running over hope like water on rocks; but in all these years, he hadn't given up on them—never said goodbye, never stopped looking. When I'd asked him if he thought the golden toads were really gone, Eladio had answered: "It is possible they're not. There are other areas in the Reserve where no one goes. There might be more."

When I thought about the golden toads myself, I closed my eyes and imagined I was Eladio—he who had seen them last and long ago, whose hope had not diminished through the years—and I wondered if my faith would ever be as strong as his. *How long would you look for something that was lost?* I asked myself. *If you loved it, and you missed it? If you thought that it was out there waiting for you, how certain would you have to be to find it? Would you need to see it in the flesh—a glimpse of it, for just a moment? Would a chisme be enough—a rumor, a whisper?*

"La esperanza es lo último que se pierde," they say in Costa Rica; the last thing that you lose is hope.

. . .

For a long time, I thought that we would not go back. In those first few weeks after coming down from Brillante in the summer of 2018, we had tried to organize an expedition to the place of Eladio's last sighting; but the rangers were busy with the aftermath of the last November's tropical storm and the landslides that had come rolling down the mountains in the later days, and we had had no choice but to retreat back down the long and winding roads into the city, board a plane, and fly back to the United States. Eladio's confession had kindled a creeping desire in my heart; a feeling that was something more than wonder and only a little short of greed. But I suspected that a part of him didn't desire to return—that to trek back into his last site of revelation and to find it empty would be a heartbreak he could not survive, a fear that haunted him like a ghost.

When I took the old maps off the wall from time to time and tapped my finger on the cerro where I imagined Eladio had seen them, Pri would shake her head and counsel, "Why don't you leave the man in peace?"

And Kyle had turned his attention toward other species, hoping to send out what he had learned about climate and disease like messages in bottles, to avail those still fleeing from the shadow of extinction. When I asked him what he thought about going back to seek the last survivors of the hidden mountain, he answered: "Do you really want to know? What if we do go back and the golden toads aren't there? Maybe it's best to let it be a mystery." And most of the time it was; I wanted to believe in it, to live with the faith and leave the garden undisturbed.

Seeking counsel, I sent an email to Karen Masters in the summer of 2020, asking her what she thought about the idea of going back—*It would be like finding El Dorado*, I had written—and she replied:

> *Here is some food for thought: Would the rediscovery of El Dorado bring desired consequences, or is it best if left to legend? Same question, but for the Golden Toad. Yes, you allude to the consequences of finding the Golden Toad, but I don't think that you really reckon with*

them, or impart a sense of tragedy—for humanity, or for Golden Toads themselves—that a rediscovery could entail: a conundrum for the discoverer, a threat for the toad, and a squandered opportunity to mourn, learn, and do better by those who heed the message of vanishing species. What does the Golden Toad gain, and—equally importantly—what does it lose, should it "reappear"?

She was echoing the doubts that chewed my bones on sleepless nights, warnings coming out from a deep well. I feared destruction working through the best intentions. I feared a recapitulation of apocalypse. It wasn't my toad, anyway, I told myself. It wasn't my story, and it wasn't my discovery. So we agreed that we would not go back; and we said it for a long time; and for a long time, we believed it.

But the mystery refused to sleep; it buzzed like a mosquito in my ear, singing of the unwalked trails and the secrets undiscovered in the deepest places of Monteverde's unmapped forests. The stories of rediscovered frogs breathed new life on the embers; resurrection had happened before, it seemed, and it might happen again. Temptation beckoned, and the shadow of grief for the lost species lit up with the vision of a eucatastrophe, a flight of triumph down the mountain, calling out: "Alive! Alive!"

In my dreams, I saw it like it was right in front of me: a hidden sanctuary on a forgotten mountain, high above the grasping hands of mortality and extinction. The woods were dark and deep and wet with the unfailing rains, smelling thick with age and untroubled solitude; black wood colored green by moss and creeping epiphytes, shining mushrooms carpeting the shadowed understory. In the twisted canopy, the sounds of birds. A place where the first water had not evaporated, tucked forever in sweet wells and pools among the hollows, where old roots twisted and bright eyes glinted from the far side of eternity. And hidden in those celestial waters, the immortal golden toads: immune to time and death and hanging on outside the boundaries of our elemental world, the very last survivors of a long-vanished kingdom.

In the end, the possibility proved too great a temptation to resist. To forsake the trail that lay hidden, leading to the last of the golden toads, felt like leaving a great labor uncompleted—a map to El Dorado never followed and left to be forgotten. I discovered that I couldn't live with the mystery. So we went back.

CHAPTER EIGHT
ON THE MOUNTAIN OF REVELATION

Cordillera de Tilarán, Costa Rica

The summer that we found out Pri was pregnant, we'd been spending our evenings looking over topo-maps of the Costa Rican highlands. The pregnancy had been glad news after a long winter, coming in with the thawing of the snow. He would be our first; I was going to be a father, Pri was going to be a mother, Kyle was going to be an uncle. Some day soon, the stories we had heard from our parents would begin to be passed on, and maybe we would even tell this story, too, the one that we were living now. *There used to be this golden toad*, I would tell my son, *and your mom and dad went looking for it.*

But how that story ended, I wasn't certain yet. We'd made the decision to go back before we knew that Pri was pregnant, after she had spoken with Eladio to remind him of the lonely mountain and the hope long hidden in the distant forest. He had thought about it for a long time before he answered, and then finally agreed to return to the place where he had last seen the golden toads in 1991—beyond Brillante, farther out along the cordillera where few boots had ever walked. We weren't certain if we had offered him transcendence or talked him into parting with his soul.

As Pri tended to our unborn child—that summer, the size of a blackberry on the vine—we studied satellite images of the cordillera and traced the hidden trails with our fingers, telling tales of our approaching journey to a fetus that could not yet recognize our voices. We took the pregnancy as a happy omen—life emerging ahead of life returning—and we looked forward to the endeavor with a sense of hope. Ultimately, we knew we would have to put our faith in luck.

Over the last ten years, the rains in Monteverde had become increasingly difficult to anticipate, but there was a small window that we were willing to bet on: We would fly into the country at the end of May in 2021 and plan to hike the winding trails to the distant mountain in the first few days of June, when the rainy season was most likely to begin. If our guess was right, we would climb up to the ridgeline as the first storms descended, water calling the secret creatures from the earth. If the rains didn't come, the expedition would be fruitless; whatever frogs existed in the highlands would lay hidden from the dry air, and we would have no hope of finding what we sought.

Hesitant at first, Kyle had finally agreed to come along. He had spent the last three years learning everything he could about the mechanisms of disease and climate change in the amphibian declines, and he had just begun to untangle himself from the mystery that I had drawn him into. He was a person who did not believe in regrets, and so for him the quest for rediscovery was fraught with peril, risking as much as it stood to gain. What finally compelled him was the question of what we might learn from a species that had beaten death; if the golden toads had survived against all odds, then they likely held the secrets of combating the scourge that had wiped out all the others. In our particular time and place, that knowledge would be as valuable as treasure.

That June was three years since we had walked with Eladio along the old Brillante trail, and in that time a new pandemic had descended on our species, moving across continents and cultures with a raving hunger. It was not unlike the pandemic that had fallen on those forests thirty years ago, pursuing the heedless frogs into hiding or extinction. The world looked different than it had when we last looked upon the cemetery of the golden toads; we were beginning to see it through the eyes of those lost frogs, helpless and hunted. We all knew people who had died gasping.

ON THE MOUNTAIN OF REVELATION

When we flew to San José and made our way up to Monteverde, we rented a room beside the old duplex where Pri and I had lived in the old days, where we could look down on the towns of Cerro Plano and Santa Elena in the warm evening light and watch the mist sweep in across the green hills. We stayed awake until the night was dark, translating the questions that I wanted to ask Eladio when we stood at the top of the mountain together, and Kyle and I watched *Jaws* on the little laptop that I had carried out from Arizona. Spielberg had been the same age when he made *Jaws* that we were now, and something about that knowledge made our task seem less unnerving—at least we would be on land. A few days before we were scheduled to set off, we decided that Pri would have to stay behind in Monteverde; between the pandemic and the unknown trails of the wild cloud forest, potential complications with her pregnancy could prove disastrous. Kyle and I watched *Jaws* again and brushed up on our Spanish.

In collaboration with the Monteverde Conservation League, we had assembled the team that would accompany Eladio on his return: Luis Solano was one of the only forest rangers who knew the area well, so he and Eladio would lead the way through the labyrinths of the distant ranges. Mark Wainwright and a UCR biologist named Gilbert Alvarado would join the group to identify and document the species we encountered in the remote forests when the rains descended.

In the midst of the logistical uncertainties and shifting plans, I was haunted by another doubt: the feeling of opening a door that cannot be closed. I couldn't help but ruminate on the question: What would I do if I found the golden toad?

For the last three years, I had asked myself that question, and I had not yet settled on an answer. Without a doubt, the rediscovery of a species lost for thirty years would be a great boon, a validation of the conservation ethic that arose in Monteverde in the wake of the golden toad's disappearance; it would bring new hope in the face of a darkening world, a beacon in the night. But it couldn't come without a price;

the survivors, long hidden, would be carried off. Their secret temple would be defiled. What if the only peril in that place was the kind that we brought with us? In my imagination, I saw a vision of collectors descending on the mountain, of a new pandemic not so dissimilar from the last, and of the second coming of extinction. Rediscovery would inevitably present the chance for another tragedy. Could we bear to lose the golden toad again? For years, I had thought of the golden toad as a holy grail, and one that we might drink from and be saved; I had asked myself what I would do if I encountered it, but I had yet to ask another important question—one that would need to be answered before the end of this, one way or another: *Whom would the grail serve?*

These were the thoughts that circled like vultures as Kyle and I embarked at last on the final stage of the quest—as we left Pri behind and bounced in the 4x4 down the steep roads from Monteverde, past the cuatro cruces, and into the highlands above Cedral, into the forest of the revelation. On the dirt roads that ran like spider webs across the Pacific slope of the Tilarán mountains, dogs slept in the middle of the streets and cows climbed the terraces in slow and careful steps above the old houses of the campesinos. I knew that we might follow any of those roads to encounter stories of resident spirits and unknown species that had not been told outside of those communities for decades, or sit at a bar beneath an old television and order a Pilsen where only the locals had ever nursed their zarpes.* But we had no time for such diversions; our path was clear, and it lay ahead of us: We were on a search for the golden toad.

• • •

Kyle and I were finishing a quick breakfast at a little soda somewhere past the cuatro cruces in the early morning. As the doña fried dry eggs

---

* Zarpe: The last drink of the night.

ON THE MOUNTAIN OF REVELATION

and gallo pinto in the kitchen, we made small talk in one-and-a-half languages through cloth masks with Luis Solano and Mark Wainwright. We were the only patrons at the little restaurant, which seemed to have been dropped from above into the middle of the Costa Rican countryside. Luis was a ranger with Monteverde's Bosque Eterno de los Niños, the largest private reserve in the nation, and his last eighteen years working with the BEN had taken him across many of the preserved forest's twenty-three thousand hectares. He knew these rural roads and remote access points like a cat knows a jungle—but even he had not been everywhere.

"There are many areas of the BEN that are pretty much impossible to get to," he conceded across a cup of black coffee at the soda while we waited for the others to arrive. "There are incredible places out there, but we haven't reached them. There are a lot of mountains, a lot of streams that you see from afar, and you wish you could go—but maybe that is not the purpose of the expedition, or you don't have time—and they remain unseen."

As we talked, a second pickup pulled off onto the dusty shoulder by the restaurant—dirty white with big black letters reading "UCR"—and Gilbert Alvarado from the University of Costa Rica tumbled out to join us for what would be our last real meal for the next few days. I had never met him before, but I knew Gilbert by reputation. He had worked with new populations of the green-eyed frog in Parque del Agua and Sal Cerro in the Cordillera Central, and in 2016 he had rediscovered the endemic red-bellied streamside frog (*Craugastor escoces*) thirty years after its last sighting; it had disappeared in 1986, just a few years earlier than the golden toad. Since then, Gilbert had identified new populations of lost tree frogs and harlequin toads in Costa Rica, and he had also spent years studying amphibian resistance to *Batrachochytrium dendrobatidis*—learning what he could from the survivors.

Mark Wainwright stood up from his plate of eggs and salty cheese and walked out to say hello to Gilbert, waving his hand in front of him to keep away the dust that the truck had kicked up in the driveway. They bumped elbows—the new salutation in the changing world—and talked in Spanish on their way back to the table. The doña brought a second plate of fruit, then wandered back into the kitchen where white rice steamed in a big pot on the stove.

Beside me, Kyle opened his bag to find his *Birds of Costa Rica* field guide, squinting out to the forest's edge to identify a flycatcher on a wire. It was his first time back in Costa Rica in three years, and a few months since we'd seen each other in the flesh. A long road lay ahead of us, and I was glad to have him with me: We would need everything we both had learned if we ever hoped to find what we were looking for.

On a wood bench in the grass, wordlessly sipping a cup of coffee and gazing up at the distant forests, sat Eladio; his face was drawn in an illegible expression and he seemed to be wandering somewhere else in thought. It was the first time he had found himself beneath the shadow of this particular mountain in exactly thirty years.

After breakfast, we filled our water bottles from the faucet at the kitchen's sink and climbed into the vehicles—Luis and Gilbert in their pickup trucks and the rest of us in the 4x4 that I had rented in San José, hoping that it could handle the uncertain mountain roads above the last towns in the foothills. We followed Luis's pickup truck into the high ranges, where the roadside cliffs dropped off in steep walls and gliding vultures eyed us doubtfully, intruders to their airy realm. Before long, the scattered houses at the roadside were swallowed by the encroaching jungle, and creeping grasses absorbed the track until it was lost somewhere underneath. In some places, the tires of the 4x4 sunk half-deep into pools of standing water; the first rains had rumbled in the day before, bringing new moisture to the thirsty land after a long dry season. Behind the monument of the cordillera, twisting like a serpent, drifted dark clouds promising a coming storm. At the foot of the mountain, Luis's pickup rolled to a stop beside a broken fence line and was still.

ON THE MOUNTAIN OF REVELATION

We climbed down from the 4x4 and gathered by the gate, where the vague shape of a slight and hidden footpath crept through the wild pasture and wound up along the slope, then was lost among the overgrowth. A silent mist was blowing in, swallowing the high canopy in a silver blanket.

"Shall we have a picture of the group before we begin?" Mark said, so we paused for a quick photograph by the vehicles before opening the gate and crossing into the long and deep grasses. Luis and Eladio in the lead, we hiked past a leaning cabin on the edge of collapse—an old ranger lodging, Luis explained, where he had once surprised a group of poachers bold enough to spend the night inside—and through blackberry thickets speckled with stinging thorns, the first shallows of a larger sea. Emerging at a level place near the top of the field, I hesitated on the verge. Before us rose a wall of forest like a great gate, canopy waving in the Caribbean breeze coming over the far side of the Divide. Then, one by one, we stepped across the threshold; the forest closed behind us like a greedy hand, and we began a long climb into the dark.

· · ·

It was two years after Frank Hensley had seen the last golden toad in Brillante that Eladio traveled up this mountain in the wet core of the rainy season and chanced upon a miracle. In 1991, he was working on a study of cloud-forest plants with a Monteverde biologist, leading a transect from the Peñas Blancas river valley across the Continental Divide to Arancibia above Miramar. Hiking far into the remote highlands southeast of Monteverde along the Tilarán cordillera, they cut new paths into the tangled elfin forests and drank from clear streams at their headwaters on isolated peaks and ridges, documenting endemic plant species for the ecological catalogs in Monteverde. As Eladio and his companions climbed onto the wind-battered elfin forests of the mountain's spine, they found hollows among the roots of trees where dark water had collected, the same sort of hidden pools that had sheltered the golden toads of Monteverde before they descended for the last time into their secret dens beneath the earth.

But it would prove to be a special day. Descending down a leaf-littered slope wet from the afternoon rains, Eladio discovered a group of shallow pools sheltered from the howling wind, speckled with the moving forms of the enduring golden toads: glimmering survivors. There were fifteen or twenty males, several females, and a mass of juveniles gathered in a small sector, hidden on the high mountain beyond the grasp of the killer that had brought doom to the toads of Monteverde. It was like walking onto a mountain that towered above death.

The 1991 discovery proved two things. First, it demonstrated that the iconic species had not disappeared entirely from the forests of northwestern Costa Rica, as had been widely believed and reported for the last two years. Second, it challenged the perception that the golden toads existed only in the observed range in Monteverde, effectively doubling the area of their known habitat overnight. But because no one in Eladio's party had hiked in with a field camera—and because they did not have the heart to carry one of the last remnants of the lost species away from its hidden refuge—the sighting was, and remains, unconfirmed.

When asked if stories of further encounters with that population have come out of the area, Eladio says that there are rumors of later sightings. "But it is just talk," he adds. "There are no pictures or documentation of any kind."

Because the local farmers in the rural areas below the mountain have not had access to the resources that would inform their descriptions, their scattered reports have been generally disregarded.

"They do comment on a little golden toad in the area with similar characteristics," Eladio says, "but it could be something else that they saw. We are not sure, honestly, because the experts never saw them. They were farmers in the area, but they knew them well. I don't know if it's the truth."

ON THE MOUNTAIN OF REVELATION

When Eladio and his party returned to Monteverde with news of their discovery, they were met with a similar uncertainty. The encounter has been recorded in a small handful of references and local publications, but Frank Hensley's final observation on Brillante in 1989 remains the most widely reported last sighting; and despite several attempts to return to the site where the small population was observed in 1991, no formal expedition to confirm the rediscovery was ever undertaken.

On that day, Eladio watched the toads with quiet fascination, silent witness to their unlikely existence; and as a blue evening began to settle over the mountain, he descended again down the paths he'd come from, leaving the golden toads to their quiet life in the enchanted forest.

It is impossible to know all the ways that mountain changed in the lost years. The paths that were cut through the tangled thickets were consumed by creeping vines and old roots grasping in and out of soil; the traveling wind pulled down heavy trees, bringing sunlight into some places and leaving shadows elsewhere; the floods washed old stones down the canyons and the routes of water were redrawn. For decades, the bright eyes of the forest would have seen no trespassers beyond the slow dantas,* the singing monkeys, the sly cats prowling over their dominions. From those ranges, the barking dogs and the purring motors would be lost among the other sounds of the forest, the world outside walled off by distance. Occasionally, the odd encroacher might have passed nearby on business of their own—a forest ranger marking the boundaries of the reserve, or a local farmer on an illicit hunt far from home—but it is unlikely that intention ever carried any human soul to those hidden pools lost among the elfin forest.

Today, Eladio's discovery is almost forgotten, thirty years between the past and present, and Eladio recalls only fragments of the memory; he will tell the story if you ask him to, but at the end of it he will always say the same thing: "I never went back."

---

* Danta: Tapir.

THE GOLDEN TOAD

• • •

It was early in the afternoon by the time we reached the first river crossing, where cold water ran among the mossy rocks in clear streams and white foam. Luis had led the way from the pasture where we'd left the trucks below, going up ahead to find the signs of the old path and disappearing from time to time into the surrounding woods, only to reemerge unexpectedly at the rear of the party before striding on again. Eladio walked with Luis for much of the way, but he often stopped to examine the orchids that grew on thin vines twisting around branches dark and laden with growth. On the lower slopes, spider monkeys had watched us from the canopy, shaking the branches; but as we climbed higher into the cavernous forest, the way felt lonely and remote.

After a few miles of slow climbing on the lower slopes, Luis had led us up a steep trail on a treacherous hillside, slick leaves slipping down into a ghastly blank below. For a few moments, we balanced on the edge above a towering drop, and here I was glad that Pri had stayed behind. Kyle and I exchanged silent looks that I recognized from the more treacherous adventures of our reckless youth. Then we found our footing and escaped the slope, pushing deeper into the forest. In other areas, we found twelve inches of mud and swampy places that threatened to swallow our rubber boots, and maybe more of us if they were hungry. Though we had climbed high on the mountain, only weak sunlight reached us through the canopy, and in many places we found it dark and difficult to see where the lost trail led, even though Luis had gone ahead to scout the way.

Through all of these difficulties, Eladio moved with the sure footing of a man thirty years younger—like he had left a strong and eager piece of himself here on his last visit, and had only now discovered it again. At seventy-three, he was the oldest in our party by approximately twenty years, and many men his age would have thought twice about a journey so distant and so punishing. But he showed no signs of slowing as the day wore on, and often he would disappear into the shadows ahead like a phantom caught between the present and the past.

When we crossed the river, we rested for a while by the water, and as the others emerged one by one to sit on the cool stones in the riverbed, I discovered that I was surrounded by fathers. I had met one of Eladio's sons five years earlier, at the old farmhouse deep in the valley of Peñas Blancas on the Atlantic slope of the cordillera. Gilbert had come with news that his wife was pregnant, and he would meet his first child a few months after our return; on the lower slopes of the mountain, Luis had shown me the tattoos of his children's names up and down his arms; and Mark had spoken of his new boy, only two years old, whom it had been difficult for him to leave behind in Monteverde. A few years earlier, Mark had lost his fifteen-year-old son in an accident, and he was still carrying the grief. I could not imagine what that would feel like.

I wondered what I might learn from these four fathers, if I studied them the way a herpetologist studies a frog. Would I learn to care for an infant that could not yet keep itself alive, brand-new to the world, counting on me? Would the generations of hereditary instinct reawaken, or would I reach for them and find that they'd abandoned me? Would I learn to be patient, to have faith in the dry years that the rains would come again?

For years now, the trail of the golden toad had been the place that I invested all of my attention. In the last few months, I had begun to fear the answer to another question: When the time came, could I learn to set aside my best ambitions to be a father?

I wasn't sure—but I had an example I could follow: After Kyle and I were born, our father had reduced his dedication to his research by degrees—he would publish a new paper here and there, exploring mimicry in butterflies—but his long days spent chasing queens and viceroys through the prairies drew slowly to an end. Instead, he spent his evenings teaching us about the small cocoons he still sheltered from the cold, and his weekends taking us to waterfalls to look for salamanders in the rocks. His children had become his great ambition; he had positioned us beyond the tide of all his other dreams. I didn't know if I would rise to that occasion when I had a child of my own.

I regretted that our father wasn't with us now, beside the singing river, so that I could ask him for the secret. As we had planned our great adventure to this mountain, he had asked me if I thought the trail would be something he could handle; it was the closest he would ever get to asking us if he could come along. I had told him that the old paths would be wet and overgrown; that the journey would be long and unforgiving; that with his back in pain the way it was, he probably couldn't make it. Sitting there beside the river, so far from home, I wished that I had answered differently. I wished that I had invited him along, so that we might have looked together on that lost world, and that he might have seen again the old forests that he had walked so long ago, before Kyle or I even existed. I wished that I'd had faith in him, the way that he'd had faith in me.

At the river, it began to rain; a few lonely drops fell from high above to join the cool vein running through the forest, and then the woods were alive with the patter of water on thick leaves: faith rewarded in a wounded land.

We left the riverbanks and climbed higher, and in the rain the green forest began to bloom. In the muddy light, deep-purple orchids glistened with reflecting water against dark leaves, and bright-red bracts of hot lips and heliconia flowers stood out among the hundred shades of green. The clouds descended into the forest and mist rose like a ghost from the understory, catching us between two worlds. As the rain fell high above, infant streams ran down beds of clay and rock, making temporary pools of black water that were so dark, they looked deeper than the ocean. Tiny leaves of ferns bent to earth from the weight of water.

The twisted trees made tunnels of wet limbs bowing low across the path, like those high on the Brillante trail back toward Monteverde on the ridge. At the top, we watched the sky turn orange, and the patter of water falling from wet leaves and long branches was all around us. We ate the last provisions we had carried and let the light wind dry our wet clothes. The light of evening began to darken, and we knew that night would soon settle over the secluded mountain.

We spent that evening searching for the golden toad. In the fading light, we spread out into the forest, each of us following a different hidden path toward the sound of trilling streams or light glimmering on water. For long stretches I was alone, and the occasional flashes of someone's headlamp—like sparks in the dark—were the only signs that the party had not disappeared, stumbling through some undetected portal into another world.

It is common to get lost in the cloud forest; I have achieved it before without difficulty. When full night descends, the darkness is not the same as it is in other places; it is deeper and thicker; it hugs you like a snake. As I pushed through heavy leaves and up into wet hillsides, I looked back to try to keep my bearings centered on the trail, but it was impossible; I would have to trust to instinct—and the murmured voices that seemed to scatter through the forest like echoes—to find my way back. I wondered which direction Kyle had ventured, and I hoped that he would not lose his way in the long dark. I felt responsible for him, but he had disappeared without my notice and was at the mercy of his own senses now, and of the forest.

While the silent wind moved through the woods, I found sleeping hummingbirds perched on fragile branches, and I slipped past, certain that their hearts would stop if I were to wake them up. Large insects, emerging from cryptic hideouts, scuttled up and down long branches, and nimble spiders sent glittering webs across the open trails, words written in their languages: *This is not your home.* In some places, the light from my headlamp reflected silver eyeshine from prowling creatures moving through the misty forest, but they crept along unseen, on their way to other errands that would not be known to us. I could only hope that they would let us pass in peace, if we were gentle and did not offend them.

By the long body of a fallen tree being consumed with moss and creeping vines, I met Mark emerging from the thicket by a small pool, his hair wet and his expression grave.

"Well," he said, "I definitely feel that now we're in the habitat. That back there, and just before it, looked like the classic spot."

Like me, Mark had arrived in Monteverde after the golden toads were already gone, but I had heard stories of him going out to search the highlands for the missing species—even neglecting important meetings on rainy evenings to hike up to the lost pools of the elfin forest in the hope that the golden toads might have reemerged. Talking with Eladio on the ridge, Mark's watchful hope had been kindled like an infant fire on a windy night.

"I always heard, and Gilbert, too, that '89 was the last year," Mark said. "And I just asked Eladio again and he said, '*No, it was definitely here; it was 1991*'—because it was the same year he got bitten by a fer-de-lance, so he remembers very well. And so I guess they were still around, at least in this one place."

We stood by the log together for a few minutes, listening to the remains of the rain dropping down from wet leaves in the canopy, and then we went off in different directions again, and Mark's light and sound disappeared behind me like the forest had swallowed him whole.

As I moved on, the woods were alive with the calls of hidden dink frogs, which drew me deeper and deeper into the thickets only to disappear, leaving me to find my way back along the closing paths alone. In the dark, it was impossible to tell where the land rose and fell, and the veil of mist caught the light and scattered it, so the same faith that had carried us here was our only guide in the labyrinth of the black forest.

I found Luis changing the batteries in his flashlight beside a little pool, where small, dark shadows fluttered in the clear water. He did not seem tired, and he was moving quickly; already I had seen him traverse the slope that the old trail cut through, waver for a few moments at the edge of the liminal world, and disappear. Now he had emerged. It was as if he was using the secret tunnels of the golden toads to move through the forest, appearing occasionally by the secret water, then vanishing. He stopped long enough to say that he had not found anything in the lower pools, and then he was off again, pushing the tangles of trees and vines aside like a curtain. As the night deepened and the search wore on, Luis—a recovering realist—began to seem more optimistic, and his eyes danced with a hope that he wouldn't say out loud: *We might actually find it.*

In the early evening, I had asked Luis what he thought about our chances. Cautiously, he had replied: "My dream—and I've always said it—is to find the golden toad." That desire is one that most of the older generation in Monteverde shares, but not everyone will say it out loud. Thirty years is a long time, and many have given up the search; and when they go into the woods these days they no longer look for the spark of orange among the fallen leaves. But when I spoke to Luis, he was more sincere than I'd expected.

"Since the first time that I came to this area," he said, "I have said to myself: *One day I am going to find it in these mountains.* And every time that I go to a place that's similar to their habitat, or near the area, the idea sticks with me like a thorn: *Luis, you have to go find it. It could be here.* But that is my dream: to find it for real, and for it to return."

And because he knew the forest so well—how many hours had he spent exploring this cordillera?—Luis's odds were better than most. So I followed the low snaps of his footsteps beyond the pool and deeper into the elfin forest, but he had already disappeared somewhere in the brown swamps ahead. After a moment, even the sound of his boots was gone.

Picking my way back down the hillside, I could see a faint white light far off, and I imagined that this must be Gilbert somewhere below, peering into crevices among the roots. I could barely see him in his ghostly light, a vague figure traced in outline, searching among the roots for

THE GOLDEN TOAD

golden toads. I began to feel superstitious—as if to speak to him would be to break an enchantment—so I stood at the boundary of his little light for a while to watch him search. Then I left him there and started walking back up the forest slope.

It was almost eight o'clock, and the sun had been gone for two hours. The wind felt restless, drifting up and down the mountain like a wanderer; and when it strayed to the ridge above, the forest was eerily still. In this reverie, I emerged from the murky woods to discover that I had found the trail, where Kyle was sitting by the bank like he'd been turned to stone, his head turned up and his ears listening. When he saw me, he asked: "What does the golden toad sound like, again?"

I tried my best to describe the calls as I had read about them—a trilled release call, or maybe a *tep-tep-tep* like wooden spoons—but of course I had never heard them myself, being of a different world. But he shook his head and said: "That's not what I was hearing."

I went to sit beside him and he told me where he had come from—farther up along the old trail and higher toward the ridge—but he had not found anything aside from empty pools and roaming insects, and strange calls moving through the dark. Happy to have found each other, we knew that we couldn't stay there forever waiting for the others; this was, after all, the one chance, the only chance, and we must continue on to confront either a temple or a tomb. Before long, the wind had found us again and it grew cold on the trail in the open, so we turned south and began to move down the steep slope on the left of the path, embarking on a final search.

On the side of the mountain, we found descending stairs of water, rainfall gathered in deep pockets that might have been a giant's footsteps in the stone, and we followed those farther down the hillside like a new trail. We discovered deeper pools hidden beneath undercuts and thick roots, where the ground turned hollow and water lay beneath our feet, like finding gold underneath the forest floor. Here, the water was clear and we could look down to the bottom, where sand and stones lay undisturbed for countless years, touched only by the roll of thunder and unseen currents in the pools. The roots of the old trees were covered in moss and ferns and other epiphytes, creeping deep into the hollows to places that our flashlights couldn't penetrate. We searched in these little caverns earnestly; they were the best places we had found on the mountain yet, the perfect sanctuaries for a last survivor flying from ruin. We imagined the lost toads tucked away and waiting to be discovered, dark eyes looking out of collapsing cellars. But they lived only in those images; the pools were still and empty.

Eventually, Kyle departed to hike back up the ridgeline, where we could see the flickering headlamps of Gilbert and Luis dancing through the clutching branches. I determined to follow the giant's steps a little deeper into the dark, thinking that the fog might break and I might stumble into the scene that I had envisioned in the witching hours of my sleepless nights: a congregation of a hundred golden toads around a gleaming pond, sacred and immortal fires lit about the foundations of eternity. As I followed the holy trail deeper, I moved more quickly, spurred by a mounting distress that I might miss them by a moment, that they might be only hiding, cloaked in shadow beneath the next buttress root, and then dip again below the ceiling of the earth; little fires going out while they fled my footsteps like one running from a rolling storm. To be so close and to miss them would be too much to bear, and suddenly I was sure I felt their dark eyes on me, and they were only waiting to be discovered. They were here, they were here, and in a few moments they would be gone again. My rubber boots splashed cold water as I dropped down to the level ground, and the mist blew off, and I walked forward to learn the answer to the mystery.

There, at last, I found Eladio. He was sitting beside the last of the deep pools at the bottom of the hillside, fog swirling around him like a spell; his back was bent with the effort of the long hike and from the burden of carrying our hope.

THE GOLDEN TOAD

I don't know how long he had been on that side of the mountain, peering into the dark, uncountable crypts with his fading light while the rest of us struggled through small trees and thickets on the ridge. He must have looked everywhere for signs of them, in the little pools and among the roots and in the dark tunnels of the earth, in his final desperation maybe calling out, *It's me, I've come back, do you remember?* in the empty valley of the long night.

High above, the clouds had blown off and a silver moon sent its enchantment washing over the forest. In the dim glow, the mystery was peeled away by fingers of illumination; there was only enough light to reveal everything. Then I understood with an inevitable knowledge: We would find no golden toads on this mountain. If they had ever been here, they were gone now.

"Que buen lugar," Eladio remarked without turning, taking in the site of last hope that he had finally returned to after thirty years, to find it utterly abandoned, "pero no hay."

What a beautiful place—but they are not here.

The lonely forest swayed in the slow wind, reaching out wet and woody hands in consolation; we were only now encountering the grief, but the forest had been living with it. I wondered what the other animals on that mountain had felt as they had watched the golden toads descend into the dark, never to return; now maybe nothing more than legend, were there any left that still remembered them? The clearings in the elfin forest where I had once imagined the survivors now seemed to be the empty rooms of a great chapel, filled with a melancholy echoing. Among the decayed holes in the mountain, faint moonlight washed over tumbled graves. We would return to that mountain and other sites in the following days, but we would find no signs of the golden toads.

Wandering somewhere below the ridgeline in the dark, I made my peace with the ghost. I had come as close to them as anyone, but they were still beyond my grasp—out there somewhere, drifting. I would surely find them in the shadow-forests of my dreams, but it was not my destiny to meet them here. I knew that I would sleep more peacefully that night and in the nights ahead of me, because I had committed to the final question, and I had ventured to the mountain for myself to look into the old reflecting pools, and I knew that when I was asked again, I would have an answer: I would say, *I have been there, and I did not see them.*

Eladio and I found the others scattered around the trail that ran through the tunnel of the stunted forest on the ridge, empty-handed and worn down from the long night searching. We paused to swap stories and to drink the last of our clean water, and then we shouldered our burdens and slipped back in to our wandering procession, heading home.

From a windy silence, as we descended in the mist and clouds, the distant calls of the night creatures drew nearer. Though the golden toads may have disappeared from the mountain after all those years, there were other frogs who had not abandoned it; on our last descent, they emerged, one by one, to peer out at the strange figures drifting past their homes like phantoms in the dark.

We must have been a strange sight to those frogs, who had lived their whole lives on that secluded mountain, wind and rain and silence their only kings. How many years had it been since they had looked on one of our kind? Had the stories of us trickled down like water, generation to generation? As we walked through their deeper sanctuaries, I wondered if they understood that we were plunderers, world-eaters, capable at any moment of carrying them off to a hopeless fate—wrapped in plastic in thieves' hands, never to be seen again. Did they know enough to tuck their young away in dark hollows as our ghostly figures drifted past?

As apprehensive as they may have been, our unfamiliar forms were not enough to stop them singing their love songs as nighttime settled over the mountain; the bravest of them emerged to perch on the long leaves of heliconias, wet bodies glinting in the dark, and chirped wary benedictions through the mist—their shy prayers for survival. In the canopy of green stems and from dark root hollows—above delicate strings of eggs—they came one by one from their hidden places for the night-gathering, a custom older than our species, handed down across long lineage.

After all, why should they have feared what they did not know? Their world was simple and immediate, undismayed by apprehensions of the future. Their world was the pool of water in the stone bowl of the high ridge—the elixir held in cupped hands—the cool mist blowing down from heaven, green leaves that climbed and fell and then decayed, burying their dead in barrows that they never visited. But the end of the world was closing in on them, whether they knew anything about it or not; they would be as anchored to that as any of us, would be at the mercy of the dry wind creeping up the mountains when the last rains failed. In those ways, we were the same.

But the end of their world was not here yet, tonight, as we descended through their havens. In the charmed air, we cut our way into a monument of high enchantment to find the last pools reflecting starlight, the nurseries of the future generations; when the white beams of our headlamps hit the water, a beguiling sort of light arose, but we could read no signs or omens in the water.

As we moved on, I watched Luis disappear into the steep runnels dropping into darkness at the edges of the trail; he had given up the search for the golden toad, but he couldn't help looking for other frogs long thought lost to these altitudes. One by one, he carried out translucent glass frogs glittering in wet mist, brilliant forest frogs and green-eyed frogs, the blessed survivors, and tucked them back among root-caverns and shelters of crypsis among the leaf litter. They were not orange, and they were not famous, but their stories were the same; they had looked into the well of oblivion and climbed back out. That night, they seemed to us like missing children roaming home.

And then there was a small commotion in the tree-ferns, where Luis had called Mark over to examine a small green frog perched above a rain pool on a heavy leaf, its body painted with a bright streak, like paint.

ON THE MOUNTAIN OF REVELATION

"Kyle, Trevor . . ." Mark called. "This is a big deal."

The frog Luis had found was called *Isthmohyla angustilineata*: the narrow-lined tree frog. In the light of my dying headlamp, I flipped my field guide open; I had never seen the frog before. Very few people ever had; there were less than 250 mature individuals of the species believed to exist in the wild. Over the next hour, we found two more; a second adult and a juvenile.

I took a photograph so that the old biologists wouldn't doubt our claim, then sat on my knees looking at the little frog. Our lights glimmered on his wet skin, and he clung to the green leaf as if it were a lost ark that had surfaced at last above apocalyptic waters. There was still a raw spot in my heart that I'd been holding for the golden toad, and I knew that grief would linger, even as my future took me far away from these enchanted mountains. But here was something unexpected, revelation in a place we hadn't thought to find it—instead of resurrection, a new hope. I began to think of how I would tell this story to my son, as the world continued changing all around us. If we, one day, found ourselves on the top of a mountain with no place left to go—on the ridgeline between life and afterlife—this frog might give us hope. In a world of ghosts, he was a survivor.

We don't always find what we are looking for, I think, but sometimes we find other things—and those are the things that deserve our care the most. For all of us—and especially for those who had lived for long enough to watch the frogs of Monteverde disappear—the encounter with *angustilineata* opened up a world of possibilities.

"The fact that we found three suggests there'd be a lot more if we had hit a heavy rain or a breeding bout," Mark said later. "If we found three, and they've been gone for thirty years, then there's a sustained population; one that is resilient to whatever calamity we may attribute the decline to."

For Gilbert, who had spent years searching for the amphibians that had disappeared in the population crash at the end of the 1980s, the encounter was a validation of an embattled faith.

"I am surprised to see the species, of course, but I am not surprised that the species are there," he said. It was like Luis had said: With so much forest to explore, it was impossible to ever really be sure of what was out there. "To be honest," added Gilbert, "I think that the *Bosque Eterno* has a lot of surprises."

Gilbert swabbed the frogs and tucked the vials away for later examination; he would carry the swabs back down to the university and test the recovering populations for the presence of chytrid. Their existence on the mountain meant that they had either escaped its bitter grasp or had developed a resistance to the deadly fungus. If they had endured its best efforts to annihilate them, those frogs might hold the answers to a way of fighting back against the pathogen and the changing climate.

In the impenetrable dark, Gilbert offered: "Maybe the survivors will tell a different story."

As we descended down the hidden trail in the dim light of our headlamps, a slow mist swept over a lively forest, and the calls of frogs faded as we left the ridgeline. We all followed Eladio, trusting him to find the way, as he had found his way back to this lost mountain after all those years. At the boundary of the forest, we lingered for a moment. The mountain loomed above us like a monument, reaching into the heavens where innumerable stars glinted against a deep blue sky. If we all create our own mythologies, then mine lived here: high on the lonely mountain, which had contained all of my hope and mystery: the Olympus of my myth. After all my searching for a ghost, I was beginning to feel a little like a ghost myself, stretched thin and fading in the dark. At last I left the forest and made the final descent—toward the warm glow of family, and the end of a long journey.

· · ·

Three days later, Kyle and I found Pri waiting for us in Monteverde at the Café Orquídeas, smiling in the sun in front of a plate of gallo pinto, and we told her all we could remember from the strange trails of our journey. She covered her eyes at our recollections of the steep cliffs and treacherous descents and put her hand on her stomach, thinking of the baby. She asked if we had found the golden toads—we wouldn't keep the secret from her, would we?—and we told her that we hadn't seen them. But we also told her what Mark Wainwright had muttered to himself the next morning at the cabin, pacing up and down the porch with a doubtful eye cocked sidelong at the distant ridge. "What I most got out of that trip, conceptually," he'd said, "is how easy it is to overlook what's there." But despite his hesitations, Mark had been relieved to be heading home to his new son, who—though he could not express it yet—had missed his father. When we showed her the photos of *angustilineata*, Pri said it was one of the most beautiful frogs that she had ever seen.

Kyle lingered in Monteverde for a few more weeks, and we walked the old trails above the station and the paths that danced a salsa with Brillante; I think a part of him still hoped that he might spy a flash of gold among the leaves, and history would be changed and the town would celebrate the rebirth of its long-mourned love, but we never spoke of such a hope out loud. By the middle of June, his time in the tropics had come to an end. Pri and I stayed with her family above Cartago and visited the old churches where the holy water dripped from secret fountains, hoping that some divinity might attach itself to us and our precious, hoped-for future.

And then twenty-one days after coming down the mountain, we went to our second ultrasound appointment in San José and found out that sometime in the last four weeks, our fetus's heart had stopped beating. Only later would I wonder if it had been the same thing that had sleeplessly hunted my father since his boyhood: the heart condition that the doctors said might end his life at any moment but which had never caught him yet. At the time, these ideas did not occur to me; I had only a vague notion that something that I loved had slipped through my fingers and disappeared, its vanishing unnoticed until the last. Pri and I cried together and held each other's hands, and told each other that he had only ever been here for a moment, that he had never really understood all that he was missing—that he had just passed along, just kept going.

Walking out, we saw that the stone walls in the parking garage of the Clinica Biblica were painted with the images of Costa Rica's most cherished animals—the spider monkeys, the coatis, the hawksbill sea turtles swimming in cerulean salt. On one tall pillar, high above, we saw the glowing visage of the golden toad looking down on us. We climbed into the car, and the city disappeared behind us as the long road drew us back into the clouds.

Afterward—after the death rituals that we shared with our families, and the ones we kept to ourselves—we traveled together to the Caribbean and the wound scarred under the warm sun on blue water, howler monkeys and great scarlet macaws singing their gospel songs from the canopy of the coastal forest. Through the long days of that summer, my thoughts rarely strayed from the memory of my son and the golden toad, strange companions treading into forests darker and deeper than the ones before me, ascending onward through the dark. I like to imagine that they are out there somewhere together, beyond the stars—voyagers on a long journey, glowing faintly but more brightly after a brief stop on the earth. They will go on up ahead of us to see what lies beyond, finding a path through dark and cold, and leave us here alone in the sweet forests to tell our ghost stories and murmur in the night: *Good luck, we will miss you, we love you.*

CHAPTER NINE

BURIAL OF THE DEAD

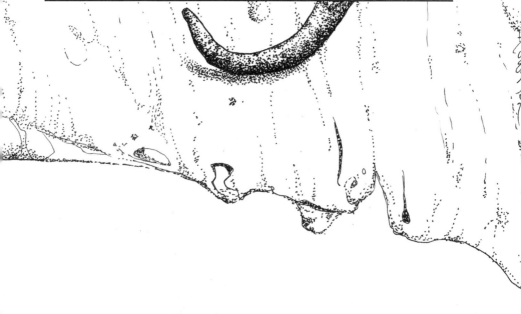

Atlanta, Georgia

Be very quiet, and very still, and if you listen closely you might hear the final song of a very lonely frog. The frog's name is Toughie—christened by the two-year-old son of his primary keeper—and he lives locked away in a biosecure facility called FrogPOD: a stout gray shipping container that looks more like an army bunker than a home. It is December 2014, and, under the care of a herpetologist named Mark Mandica, Toughie endures empty day after empty night, as the years slip by like field mice, waiting for companionship that will never come: He is the very last of his species known to exist on planet Earth.

Toughie had not always been alone. Once upon a time, he had lived with other Rabbs' fringe-limbed tree frogs (*Ecnomiohyla rabborum*) at Zoo Atlanta and the Atlanta Botanical Garden, the last refugees of a cataclysm that had ended their world. But deprived of their wild and free existence, the frogs began to wither and soon faced an approaching annihilation: Since their captivity began, the adult frogs had been reluctant to mate; and when they did, their tadpoles proved difficult to manage. No offspring had ever been recorded to survive. One by one, oblivion claimed the frogs. In 2009, the last known female of the species passed away. In 2012, the health of one of the last two males began to rapidly decline, and the decision was made to euthanize him—both to preserve genetic samples for future researchers, and to spare the frog its suffering. The Zoo's deputy director was heard to remark of the decision: "It is a disturbing experience, and we are all poorer for it."

THE GOLDEN TOAD

And then there was Toughie, an emblem for conservation and extinction, and the very last of his kind. He was featured in news stories and documentary films, his image blown up into posters and projected on the sides of buildings. But he lived out his days in loneliness. Without a mate for Toughie, it was only a matter of time before the species flickered out.

"One morning," Joseph Mendelson, then Curator of Herpetology at Zoo Atlanta, wrote in 2011, "I will find the male dead of natural causes . . . and then I will preserve the specimen for the museum shelf . . . and proceed to describe another new species of frog that used to exist somewhere in the world."

If his life had been allowed to play out as his ancestors' had, Toughie would have spent his wild days high in Panama's forest canopy, a brown, stocky, wide-eyed frog leaping among the limbs and gliding through the air on the outstretched webbed skin of his toes. He would have found a mate by calling for her from the pocket of a tree where water had collected; he would have guarded her eggs and raised the tadpoles, allowing his children to scrape off small flecks of his own skin to feed on, giving himself away piece by piece for the endurance of his lineage.

Instead, he sat on his log, in his bunker, alone. Toughie had stopped calling for a mate shortly after his collection in 2005, and the call of his species had taken on an almost mythic status—few had ever heard it, and it was assumed by many that it would never be heard again.

Then, one day in December 2014, Toughie called again. It is impossible to know for sure what compelled him to let that song out in the dark, or what it meant to him. Maybe he suspected, in the marrow of his bones, that he was the last of his kind to walk the earth. Maybe he could feel the deep ancestral beckoning to pass along everything that he had ever known and felt. It could have been his last soliloquy for a future that had been taken from him, or a mournful reverie for the world that he'd been taken from. Or was it merely an instinct, rising from the mire of his memory, echoing out into the night? A lost call for a mate—a companion—a frog to spend his final days with. We cannot know the answers to these questions. But one thing we can gather from the call, I think, is this: He still had hope, even at the end.

Toughie died sometime in the early morning of September 26, 2016. Another species vanished from the earth—its ghost going off to wherever the ghosts end up, where the forests are beyond the reach of loss and sorrow, and the nights are full of frog songs.

In the years that followed, Toughie became a symbol of extinction. But in his life, he had stood for something else: the last lingering emblem of a desperate act that had taken place nearly ten years before his death, an attempt to stand against extinction—an act that marked a shift in our collective efforts to save amphibians from annihilation, and that has set the stage for what's to come.

El Valle de Antón, Panama

Joseph Mendelson and Ron Gagliardo were weaving their way through the crowded airport in Panama City when they heard the shouts from the officers behind them. They froze. In their arms, they held large suitcases as delicately as undetonated bombs, and they had made it through the security checkpoint by virtue of a boon that Mendelson carried in the pocket closest to his heart. It had all seemed too easy, of course, when you considered the cargo that they carried.

The guard caught up with them just before they reached the gate, and he spoke the words they had been dreading: "Por favor," he said, "abran sus maletas."

Mendelson lowered his suitcase gently to the ground; but instead of opening it, he reached into his breast pocket for a folded piece of paper and handed it to the guard. It was a letter signed by the president of the airline, instructing that the travel of the two American biologists should go unhindered, and that the cargo they carried should pass unmolested.

But the guard shook his head, appearing confused, and handed the paper back to Mendelson.

"Por favor," he said again, "abran sus maletas."

Ah, shit, thought Mendelson. *Here we go.*

Mendelson and Gagliardo exchanged a hopeless glance—they had no more tricks up their sleeves—and slowly unzipped their bags. Inside, the rolling suitcases had been gutted, linings and dividers ripped out and replaced by structural Styrofoam—along with dozens of plastic containers, their bottoms lined with moss. As the eyes of other passing travelers began to wander toward the confrontation, the guard reached out and tapped his finger against the side of a container.

"Abra esta," he instructed.

No, Mendelson thought. *Any one but that one.*

But the hard eyes of the guard allowed no argument. Hesitantly, Mendelson knelt down and—doing his best to cover the gap with his hand—peeled off the lid of the container that was home to *Pristimantis caryophyllaceus*—informally nicknamed "the rocket frog."

The frog was gone in an instant, shooting off into the crowd and sending scattered cries and exclamations from the passersby; Mendelson and Gagliardo sprang up after it. It was a precious quarry, after all, and they could not bear to lose it when it had come so far—out of the very jaws of death.

Joseph Mendelson and Ron Gagliardo had come to Panama as a result of Karen Lips—Mendelson, an old friend of Lips's, had heard about her unpublished findings from El Copé and had helped to call together a herpetological brain trust at Zoo Atlanta, where he worked as the curator of herpetology. Those who heeded the call had begun to feel collectively frustrated and helpless in the face of *Batrachochytrium dendrobatidis*—by then recognized as a global amphibian crisis. The new data from Karen Lips indicated a directional spread of chytrid through Central America—from Monteverde, to Las Tablas, to Fortuna, to El Copé . . . and beyond. "Beyond" was what their group had come to contemplate. The creeping fear had been caught at the scene of the crime, unmasked, and identified—but it hadn't been stopped. It was still out there, killing at will, and it was moving all the while into increasingly vulnerable amphibian communities. What was to be done?

Once again, the zoologist George Rabb compelled the group to action. It was Rabb who had initiated the 1997 gathering of researchers from the United States, Central America, and Australia at the University of Illinois, which had resulted in the discovery of *Bd*. Now, seven years later, many of his contemporaries—Joseph Mendelson,[*] for one—considered Rabb to be the spiritual and intellectual leader of the herpetologists at war with chytrid. At the Zoo Atlanta meeting, Rabb had wasted no time assigning tasks and missions to those who had assembled.

The task for Karen Lips? *Publish your findings from El Copé, as quickly as possible.*

For Ron Gagliardo and Joseph Mendelson? Rabb's eyes fell to the map of Central America, and the projected path of the chytrid's spread. *Get ahead of it.*

Like Karen Lips at El Copé, Mendelson was tasked with anticipating the arrival of *Bd* in an unaffected ecosystem; but this time, the intent was not to study, but to intervene. His would be a rescue mission.

The concept of human intervention in ecology has a spiny history in the annals of conservation. On one end of the spectrum are those who would advocate to allow natural cycles and interactions to occur without human interference—even if, in some cases, those natural factors brought an ecosystem to the brink of collapse. On the other side of the spectrum are the conservationists who consider the protection of an ecosystem, a species, or even an individual animal to be worth the disturbance of human intervention. These were the considerations that

---

[*] Mendelson would later name a species discovered in El Valle after Rabb and his wife, to honor their efforts in amphibian conservation. The most famous Rabbs' fringe-limbed tree frog would, of course, be Toughie.

THE GOLDEN TOAD

ran through Mendelson's mind when Rabb charged him with his assignment. From his work with Zoo Atlanta, Mendelson had experience with captive breeding and *ex situ* conservation, but it was still a controversial topic. Conservationists generally agreed that *in situ* (literally: "in-place") conservation was the most appropriate response for threatened species.

In situ conservation involves addressing and strengthening the health of the local ecosystem and species populations within the context of their natural environment: establishing a protected habitat, removing an invasive predator, or passing legal protections for an endangered species.

Ex situ conservation, on the other hand, involves *removing* threatened species from their natural habitat. This could occur when a population's ecosystem proves unfit to sustain them, or when the numbers of a wild population dip so low that they become functionally extinct—when captive breeding is their only hope for survival. *Ex situ* conservation efforts are often last-resort approaches, when *in situ* efforts have failed, and conservationists must face a choice between intervention and extinction.

Joseph Mendelson knew all of this firsthand, and he knew the risks that came with *ex situ* conservation. At Zoo Atlanta, he had watched some animals removed from their natural habitats become lethargic or unhealthy; fluent field knowledge did not always translate to a captive setting, and the specimens often proved difficult to care for and hard to keep alive. Breeding could be tricky to induce; and, despite great effort and the best intentions, the captive individuals could still become the last of their kind.

But for Mendelson and his colleagues, the list of choices had grown thin. They had watched what happened at Las Tablas, Fortuna, El Copé, and other sites on other continents. And while the chytrid was a natural fungus, its spread and proliferation had been anything but natural, enabled by human movement, trade, and climate change. Mendelson knew that if there was ever going to be an ultimate solution to *Batrachochytrium dendrobatidis*, it would have to take place on the ground in the infected ecosystems, *in situ*, and it would require a holistic confrontation with each of the factors that had enabled its proliferation. This would not be that. This would be a tourniquet above an open wound, a last gambit in the face of total decimation. But it was something, at least. It was trying.

When Mendelson arrived in El Valle de Antón, nestled in the crater of a sleeping volcano in central Panama, he was staggered by the sheer abundance of frogs and toads. It was obvious how the nearby Thousand Frog Stream had drawn its name: The diversity of species and the aggregations of individuals were unlike anything he had ever seen. It wasn't until that moment that he fully understood how deeply the chytrid had impacted his old field sites in Mexico before he'd reached them. For the first time, he was encountering an intact and undisturbed amphibian ecosystem—a forest as the others once had been, a lost world, paradise.

At El Valle, Mendelson and Gagliardo rendezvoused with the Panamanian biologists Edgardo Griffith and Heidi Ross, local collaborators who would prove to be invaluable to the rescue operation.

At nightfall, the biologists set off into the chirping forests to scour the understory for frogs and toads, capturing all they could and spiriting them away. They felt like poachers, in a way, or rogue collectors—but they were working for the benefit of the frogs, whether or not the animals understood the motives of the encroaching titans. The biologists had previously collaborated on an inventory of the species that they deemed most in need of rescue, which meant that many other frogs would be left behind—they were not rare enough, not revered enough by science, to warrant rescue. But they were no less alive than the others, and they didn't want to die. In the dark nights beneath the cold light of the moon, many frogs were left behind, waiting by the riverbanks for a salvation that would never come. Mendelson struggled to align the expedition's objectives with the enchanted world that lay before him: Each and every frog that he passed by, he was leaving for the encroaching Armageddon.

Mendelson thought: *This place is doomed.*

At first, the group had planned to house the displaced frogs in a captive breeding facility on-site within the boundaries of the reserve, but construction of the new facility was delayed, and the killer was moving closer by the day. In a desperate bid to remove the frogs from its path, George Rabb secured permission for Joseph Mendelson and Ron Gagliardo to transport the frogs by air, heading for an altered destination: Zoo Atlanta and the Atlanta Botanical Garden. It was in Atlanta that their wild ark would drift ashore, to unload its precious cargo.

That is how, in June 2005, the two biologists found themselves chasing a rocket frog through the Tocumen International Airport, tourists scattering and fleeing from the leaping refugee—a little green frog that carried the burden of heredity within its genome. After a short pursuit, Mendelson and Gagliardo managed to capture the frog and return it to its place among the others in their suitcases. A few more hasty explanations and pacifying appeals in Spanish to the security guard brokered passage through the terminal, and at last they sat aboard the plane that would take them home. Home for them, but a strange new world for the passengers they carried, so far from the others they had left behind. But there was nothing to be done for them now—behind them, at El Valle de Antón, frogs had begun to test positive for chytrid. These exiles had escaped in the nick of time, and there would not be a return journey: All the hope they had, they carried with them, far away.

"No one knows their secrets," Mendelson reflected later. He knew that it would be a long road before they learned to keep the frogs alive in captivity, so greatly sundered from the forests of their ancestry. "And every frog has a secret."

At Zoo Atlanta, they arranged the captive breeding populations—inventing new approaches as they went along, and hoping for the best. Hope had brought them this far, after all.

Heidi Ross and Edgardo Griffith would stay behind at El Valle de Antón. Shortly after the departure of the Americans, the Panamanian team—with support from the Houston Zoo—finished construction on the El Valle Amphibian Conservation Center: EVACC. The facility was the first of its kind, a triage site for the region's imperiled amphibians that offered a safe haven as chytrid descended on the forests of central Panama. Hundreds of frogs who had missed the departing ark to the United States were transferred from the guest rooms of the Hotel Campestre to new tanks and terrariums at EVACC. Soon, the walls were lined with lemur leaf frogs (*Hylomantis lemur*) and ghost glass frogs (*Centrolene ilex*), and dozens of other species that had narrowly escaped extinction. In a special place of honor, they set *Atelopus zeteki*: the Panamanian golden frog. It was a spiritual twin to the lost golden toad, another frog of myth. Jay Savage had once written about the species, a few years after striking gold in Monteverde:

> *The story tellers record many men who have scaled the highest peaks and searched the darkest forests for even a glimpse of the golden frog, but only a few ever see it. Fewer still capture the cherished creature and hold him for a few moments, and a very few are able to carry him with them for a longer period of time. One story tells of a man who found the frog, captured it, but then let it go because he did not recognize happiness when he had it; another released the frog because he found happiness too painful.*

For the local biologists who had swept the golden frogs out of the path of ruin, the survival of the species was reward enough, and they looked forward to a day when they might open their hands to release the frogs of myth again.

By January 2007, chytrid had traversed the Panama Canal, deflating hopes that the barrier would slow the spread of the disease to the rich forests that lay beyond. Now more than ever, the rescue operation at El

Valle de Antón proved a prescient move made at the final hour. In this new battle with a relentless enemy, the old policies were out of date; *ex situ* conservation—once considered unethical and disruptive—had become an essential strategy in the quest to save the frogs.

Mendelson and Gagliardo's frog evacuation at El Valle de Antón was not the only labor to come out of the meeting at Zoo Atlanta. Other biologists in attendance founded the Amphibian Ark, a collaborative international organization that tasked itself with the rescue and *ex situ* preservation of five hundred amphibian species from areas where *in situ* conservation offered little hope. In the same way that the rescue operation at El Valle de Antón was seen as a short-term solution in the face of an incoming calamity, the founders of the Amphibian Ark bore no illusions that their efforts would solve the amphibian crisis—it was merely a step in the right direction.

"It is not the goal of AArk's programs to collect animals from the wild purely for exhibit in zoos or aquariums," the organization states. "In fact, that is the last thing we want, and as an end point, it would represent complete failure of the program. . . . Our vision is the world's amphibians thriving in nature, and our mission is ensuring the survival and diversity of amphibian species focusing on those that cannot currently be safe-guarded in their natural environments."

In 2006, Joseph Mendelson, George Rabb, David Wake, Karen Lips, Martha Crump, and Federico Bolaños teamed up with forty-four other biologists to pen a manifesto in *Science*. Desperate though they may be, they wrote, "*ex situ* programs may be the only option to avoid extinction for many species."

"A bunch of frogs in glass boxes is not an acceptable outcome to me," Joseph Mendelson reflects today. "But I'd rather have frogs in glass boxes than no frogs at all."

These safeguard colonies for captive breeding are not a solution, but they are an attempt to ensure that fewer species vanish in the dark before anyone has noticed. They are one attempt to heed the warning of the golden toad.

. . .

Joseph Mendelson never considered the frog evacuation of El Valle de Antón to be a solution to chytrid; to him, it was just a desperate act to save a forest full of frogs from mass annihilation. He had been around enough to know that stopping *Bd* was a task beyond his means, but he had not succumbed to the despair: He had done the next good thing that he could see—he had kept walking up the path, even though the night had grown dark and lonelier.

It is an example that we might keep in mind, as we consider the current state of chytrid and a changing climate—because the crisis isn't over. Today, it is understood that *Batrachochytrium dendrobatidis* has been responsible for the calamitous decline of over five hundred amphibian species—and the total extinction of nearly a hundred. Never before in recorded history has a disease struck so terribly, so silently, spiriting away entire species before we even knew of its existence. And in the years since its discovery, the war against the chytrid has only grown more complex. *Batrachochytrium dendrobatidis*, after all, is a living being, too, working as hard as it can to survive amid the tumults of an existence on the earth—the same thing it has been doing for two and a half billion years. It continues to evolve, and its victims scramble to outrun it, or to seek a sanctuary in the deeps of tangled roots and earth. Scientists are still discovering new lineages of *Bd* around the globe, splintered siblings setting out on their own mad crusades, adapting as they go along; and each of those new variants has the potential to decimate the forests to which it spreads. There are also younger hybrids, recombinant strains of *Bd*, forged through the union of their ravenous progenitors and levying new assaults on unsuspecting victims: Amphibians that have adapted to preexisting strains have no defenses to the hybrids. In 2018, a Brazilian hybrid was found to be even deadlier than the global panzootic lineage, and a 2023 study confirmed that amphibian declines are more severe in regions afflicted by these hybrids. Hybridization will continue as distinct strains disseminate across the globe—new threats waiting for a chance to spread.

These opportunities to disperse are becoming more numerous as the climate continues to change in *Bd*'s favor. Global temperature shifts are bringing more areas into the preferred temperature boundaries of the fungus and pushing amphibians to the outer limits of their ranges: Stressed by the climatic shifts and compelled migrations, amphibians become more vulnerable to infection. Recent studies have also indicated that temperature variability can make it more difficult for frogs and toads to fight off chytrid infections. They cannot escape the touch of the pandemic, and they cannot survive the sickness. As the global climate continues to destabilize, we can expect to see chytrid's mobility and lethality grow stronger, as amphibians flee deeper into their final refuges for shelter.

But these last refuges are also disappearing, and many species are at risk of finding themselves—like the golden toads—surrounded on all sides and isolated on their lonely ridgelines, with no more mountain left to climb. The survivable range of amphibians is shrinking. Climate change is largely to blame; most amphibians evolved to survive in relatively narrow temperature ranges, and global temperature shifts are forcing populations to flee from the forests of their ancestors to find survivable conditions. They can only go so far.

Thirty years after Alan Pounds and his collaborators first implicated climate in the golden toad's extinction, mounting research has continued to vindicate their hypotheses that were disparaged for decades, and climate change is recognized today as an existential threat to the world's remaining amphibians—and to us. In 2022, the United Nations' Intergovernmental Panel on Climate Change cited the golden toad as one of two "climate change associated global species extinctions to date." These confirmations have come slowly, over time. Over the long years of contention and debate, the biologists sounding the alarm on climate change have not abandoned their watch.

"In the popular press, you still hear that chytrids wiped out amphibians, and climate change isn't always mentioned," Karen Masters reflects today. "Or the complexity of forces at play is not mentioned. And that's because the public can handle just about one factor. That's okay; I get it. And so it's going to be a while before people talk about the chytrid *and* the climate change. And then it's going to be a little while longer before it's reported that climate change manifests itself in many ways, like pathogen outbreaks."

Batrachochytrium dendrobatidis did not act alone, but with two primary accomplices. Climate is one, and we are the other. As pathogens evolve and global climate change reshapes vulnerable ecosystems, human development continues to shrink the livable ranges for amphibians. Habitat destruction and degradation, the results of deforestation for agriculture, timber harvesting, and infrastructure, are destroying the great swaths of land in which amphibians have made their homes for eons. Populations are cut off from one another as highways carve through forests, fragmenting previously intact habitats; wetlands are filled in and paved over to create parking lots; streams are polluted by industrial runoff. And when they set out in search of safer lands, it is a changing climate that hems them in.

Today, *Batrachochytrium dendrobatidis* is beginning to return to the scenes of its earliest crimes. In June 2022, two Australian scientists published a plea in *The Conversation* for readers to send in photos and observations of dead frogs on the continent. They had recently begun receiving reports of strange behavior—thin and thirsty frogs out in the open during the day, on doorsteps, footpaths, and highways, seeking moisture in plant pots or shimmering pet bowls. Forced into lockdown by the coronavirus pandemic, the Australian scientists had relied on community reports to understand the severity of what was occurring. "Across Australia," they reported, "a remarkable 1,600 people reported finding sick or dead frogs. Each report often described dozens of dead frogs, making the grim tally in the thousands." While studies are still underway to clarify the details of these cyclical die-offs, the researchers have acknowledged that *Bd* "is certainly involved." By 2022, almost forty-five years after the modern understanding of *Bd* began in Australia, the continent should have been safe—but it wasn't. The frogs were dying again.

The same might be true on the African continent. When scientific detective work caught up with chytrid's deadly trail in the 1990s and early 2000s, African amphibians seemed to have been largely spared from the onslaught—they had existed for so long beside the African lineage that many scientists once theorized that southern Africa had been the origin of chytrid's emergence. But a 2023 study recently revealed that *Bd* has expanded its foothold in Africa since the year 2000. One of the authors of the study, Vance Vredenburg, reports: "This rapid surge may signal that disease-driven declines and extinctions of amphibians may already be occurring in Africa without anyone knowing about it." What precipitated the outbreak isn't certain, but the frogs are dying like it's thirty years ago.

This sort of enduring proliferation of *Bd*, unpredicted and unexplained, casts a shadow over the forests that have remained untouched by the disease. For years, the greatest concerns were afforded to the island of Madagascar, another bastion of biodiversity and endemic species. What destruction, biologists worried, would be levied against that isolated jewel should chytrid make landfall on its shores? Then, in 2014, researchers discovered evidence of chytrid in Madagascar frogs shipped to the United States in the pet trade. They would soon find that the pathogen had been present on the island since 2010. Another kingdom had fallen.

Today, Papua New Guinea is one of the last strongholds: Home to 6 percent of the world's frog species, its islands are—at the time of this writing—untouched by chytrid. But many wonder how much longer that can last. "I dread the day," Karen Lips wrote in 2013, "when I hear that an infected frog has been found on Madagascar or Papua New Guinea." Already, half of her fears have been delivered.

One of the biologists who had been there to witness the discovery of chytrid on Madagascar's shores was a Ph.D. student named Jonathan Kolby. In 2016, two years after identifying the presence of chytrid in exported amphibians from Madagascar, Kolby launched an ambitious plan to replicate the success of Panama's EVACC facility in the cloud forests of Cusuco National Park in Honduras. Lacking institutional support, Kolby organized an Indiegogo campaign, offering incentives like photo prints, tote bags, and Honduran coffee in exchange for donations from the general public. The campaign raised over $25,000, and Kolby and his team applied the proceeds to purchase repurposed shipping containers that would become the Honduras Amphibian Rescue and Conservation Center. Soon afterward, they began a captive breeding program for the region's endangered frogs; without this intervention, many of those species would be gone today. But captive breeding was only their secondary goal. Their primary focus was on something more audacious: the capture, treatment, and release into the wild of frogs that had been infected with the chytrid fungus.

Kolby's strategy—taking the fight back to chytrid's doorstep—was one that had occurred to others as well. Beginning in 2009, a Spanish ecologist named Jaime Bosch began rappelling down Mallorcan cliffs each fall with a bag of tadpoles on his back. The island of Mallorca is close to forty-five miles long and sixty miles wide, and home to just a single species of amphibian: the Mallorcan midwife toad (*Alytes muletensis*). There is a finite number of ponds that dry out each summer and refill in the autumn; it had occurred to Bosch that, in theory, the water infected with *Batrachochytrium dendrobatidis* would disappear on a seasonal cycle—it might be possible to eradicate all traces of infection on the island. To reach the secluded ponds, Bosch rappelled down steep cliff faces to collect every single tadpole in plastic water bottles. After hiking out, he would bathe the toads in an antifungal solution to kill the infection, helicopter back to the site, rappel back down, and release the tadpoles—then he would repeat it all again at the next pond. When reinfections persisted, Bosch and his team upped the ante and disinfected the entire island, draining ponds and scrubbing the rocks clean of the infection. In November 2015, the team published their report: *Batrachochytrium dendrobatidis* had been entirely eliminated from the island. The midwife toad was saved.

The world is not an island, of course, and the Mallorcan approach cannot be replicated in larger areas—but that doesn't mean that it wasn't worth doing. Bosch's intentions reflect the sentiments of George Rabb, calling for the rescue of the frogs at El Valle de Antón: "We may not be able to save every frog, but we can sure as hell save *some* frogs." In many cases, all they need is a fighting chance.

The battle against chytrid began gaining ground as early as 2001, when Don Nichols reported partial success in treating infected frogs with antimicrobial drugs—a process that had been on his team's mind ever since their initial discovery of the dying poison arrow frogs at the National Zoo. Since then, scientists' ability to detect and treat chytrid has continued to improve. Lee Berger's lab in Australia developed an initial detection swab, but it was confounded by the new lineages of *Bd* discovered in more recent years; this was remediated in 2023 with the development of a qPCR test—the same kind of test that was dispatched to detect the human coronavirus pandemic. And the ability to reliably detect the pathogen has enabled scientists to develop more effective treatments: A 2014 study found that an antifungal drug called Nikkomycin Z can act as a chitin synthase inhibitor, slowing down the growth of *Bd* in amphibians. In 2021 and 2023, two distinct investigations found promising results in halting the spread of the pathogen and increasing resistance to chytrid in captive specimens—essentially inoculating amphibians against infection, like a vaccine. While the logistics of immunizing frogs in the wild would be challenging, some scientists are hopeful that doing so could result in a herd immunity, creating future generations naturally resistant to chytrid.

Biologists in Monteverde have observed a natural resistance to the pathogen in the green-eyed frogs (*Lithobates vibicarius*) that have reemerged in the swamps on the Atlantic slope. "Researchers went in after our staff rediscovered it and actually swabbed the frogs, and they tested positive for chytrid," says Lindsay Stallcup, executive director at the Monteverde Conservation League. "So that means that they didn't escape—they persisted." The more that local biologists can learn about these kinds of natural defenses, the more tools they have for protecting the vulnerable species still at large in the cloud forests. What we might learn from a narrow-lined tree frog (*Isthmohyla angustilineata*) discovered on a distant ridge remains to be seen.

Natural resistance appears to be developing in other areas too. Studies in 2014, 2019, and 2021 all presented compelling evidence that amphibians are capable of adapting resilience to *Batrachochytrium dendrobatidis*—in some cases through the use of other locally adapted skin bacteria to fight the pathogen.

Batrachochytrium dendrobatidis became the scourge that it is today over epochs of evolution, enabled by a changing climate and human intercession. It may not be too much to hope that, if given a chance, the frogs might find their footing in the new and dangerous world.

"I've gotten in a lot of trouble for saying this," Joseph Mendelson will tell you today, "but I'll say it again: More frogs have saved themselves than all of our human efforts against chytrid."

Mendelson isn't the only one keeping an eye on the dark patches of the forest, hoping to see the frogs and toads emerge again as survivors. In the spring of 2024, Kyle received an email from Greg Czechura, one of the Australian biologists who had been among the first to sound the alarm of disappearing amphibians; he had sent links to two news stories. The first referenced the 2018 rediscovery of the horned marsupial frog; the second, from April 2024, reported that the critically endangered northern corroboree frog had been sighted in Australia's Namadgi National Park for the first time in five years.

"It is discoveries like this," Czechura wrote in his email, "that keep me cautiously hopeful for the southern gastric brooding frog."

"And I hope we are wrong about the golden toad," Mendelson reflects, "and that really it is simply underground, waiting out this entire ordeal."

But if there are survivors—waiting to emerge from their shelters after the storms have passed—they would have to reckon with a dangerous world. There could never be a true return—a resurrection—until their lands are made safe again. We can move on from the mystery of the golden toad, but its killers are still at large, and their methods are only understood in fragments. The interactions between disease and climate are complex—always evolving, countering defenses. The frogs and toads might have been the first to fall beneath the shadow of that dark confederation, but they will not be the last. Unless we can learn from their vanishing in the mists of Monteverde, one day we will have to face the same oblivion that came looking for the golden toad. Already, others have begun to disappear.

Bunderbos, Netherlands

The first smoke signals had risen out of the steep slope forest of Bunderbos in the Netherlands. In the nearby village of Bunde, nobody stopped to look or mourn. Business carried on as usual; shoppers strolled through the central square, a pair of bikes passed by, a truck bounced across the uneven cobblestones of the old street. There was no sound of dying flames taking a last taste of the undergrowth, but a fire was disappearing all the same: The light of the fire salamanders was going out.

BURIAL OF THE DEAD

In the Bunderbos forest, among the pools and the spring blooms, the species had once been a common sight: small black salamanders adorned with stripes and spots of yellow, red, and orange, looking like warm embers beside the dark waters. But already the fire salamanders (*Salamandra salamandra*) had been pushed to the very edge of their distribution in the Netherlands, losing 57 percent of their range to habitat destruction since 1950. By the early 2000s, they were found only in a few isolated sections of old-growth deciduous forest in the far south of the country, in the province of Limburg. One of their last strongholds is the brook valley of Bunderbos; the country's largest population is now confined to three square miles of woodlands. To the east in Vijlenerbos, a smaller population lives in two square miles of brooks in the valley of the Geul river. The third population, non-native and introduced, is found in Putberg near the city of Heerlen. In each of these three sites, the fire salamanders seem to occur in scattered patches, and each site is cut off from the others by geography and development, making movement between the sites impossible.

Despite the sundering of these populations, for a long time the fire salamanders of the Netherlands appeared to be secure. Monitored since 1971, the Bunderbos salamanders had yielded high estimates of population densities: Biologists counted several hundred individuals, each with a lifespan of twenty years or more. The story of fire salamanders in the Netherlands was, in many ways, a reminder of nature's proclivity to persist in the face of human disturbance. But that all changed in 2008 when biologists began to receive inauspicious reports: dead fire salamanders scattered across the footpaths in the light of day. The scenes were hauntingly reminiscent of the stories of dying frogs and toads. Local biologists attempted necropsies, but the specimens proved too decomposed for extensive investigation, and no conclusive evidence could be gathered to identify the cause of their demise. The most likely cause was identified as "either an infectious agent or intoxication." Scientists had known for some time that *Batrachochytrium dendrobatidis* could infect and kill salamanders as well as frogs, but it had never done so with such ferocity; fears multiplied in the uncertainty, and they began to worry that the fire salamanders could be the first evidence of the killer's evolving appetite.

The steep declines grew more alarming over time. As the years went on, surveys turned up fewer and fewer healthy salamanders, and more and more corpses; the number of fire salamanders observed across all known populations flitted between 241 and 10. In 2011, despite 26 different visits to Bunderbos, volunteers only managed to turn up 4 salamanders. At Vijlenerbos in 2010, 57 visits found none.

"The decline we describe," wrote ecologist Annemarieke Spitzen-van der Sluijs and wildlife disease vet An Martel, the authors of the initial 2013 report, "strongly resembles the population crashes after entry and subsequent build-up of *B. dendrobatidis* infections. . . . However, we did not find any trace of *B. dendrobatidis* in any of the fire salamanders sampled."

As the unanswered questions piled up and the danger to the remaining salamanders mounted—populations had fallen to 4 percent of their previous state—conservationists hurried to initiate a captive breeding program with the thirty-nine remaining fire salamanders they were able to locate in the wild. They watched half of them perish in captivity at the hands of the unknown killer. For the scientists on the trail of the disappearing salamanders, it was all beginning to feel a little too familiar—that old creeping fear emerging again from the tombs of oblivion.

As An Martel performed necropsies on the more recent specimens, she noticed microscopic fungal growth in skin lesions typical to the known chytrid fungus. "Because the fungal organisms looked very similar to *B. dendrobatidis* but the tests for it were negative," she told *American Scientist*, "we knew that this was a new species."

Then it was not the same old hungry killer, but a copycat, a double—a dark reflection from the same ancient lineage. They had discovered the identity of the killer, but the scientists on its trail did not know enough to stop it; the infected salamanders under observation developed lesions and ulcerations on their skin; over seven days, they descended into a spiral of anorexia, apathy, and ataxia; then they died. Surrounded by the bodies of its victims, Martel and Spitzen-van der Sluijs named the novel fungus "*Batrachochytrium salamandrivorans*": the devourer of salamanders.

When the news broke of their discovery, a pall fell over the international community of biologists and naturalists. It had been thirty years now since they had watched the world's frogs begin to disappear, slipping slowly and inexorably into the gravity well of extinction. They couldn't help but fear that the world's salamanders would be doomed to the same fate.

In December 2013, the next outbreak was detected in Belgium, thirty kilometers from the original site. By April, the fungus had moved another thirty kilometers. "If the disease continues to progress at the same rate," Martel told *American Scientist*, "then in about twenty-five to fifty years, all the salamanders in Europe will be affected."

Wary eyes turned to the Appalachian Mountains in the southeastern United States: home to the highest biodiversity of salamanders in the world. Kyle and I had spent our early years exploring those streams and waterfalls with our parents; now the treasures of our childhood were falling under the menace of a greedy hand, reaching long and deadly fingers into the waters of the Blue Ridge. It seemed strange and cruel that we had traveled so far to peer into the shadows of the golden toad's extinction, only for a similar apocalypse to descend to threaten the places we had come from. We had left our home to follow an exotic obsession, and left behind the things we loved the most—all in search of gold.

If we had any consolation, it was this: We were not the first to be lured by the spell of El Dorado.

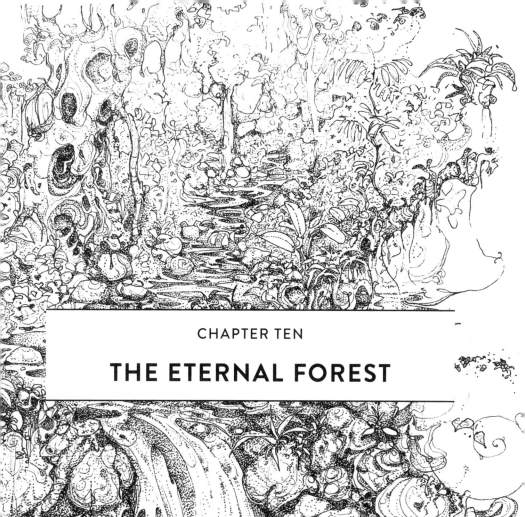

CHAPTER TEN
THE ETERNAL FOREST

Flagstaff, Arizona

The search for El Dorado exacts a price, and any who has ever undertaken it has been confronted with a question: What are you prepared to lose to find the treasure that you seek?

Pri and I came back to the monsoons and thunderstorms of Northern Arizona in August, watching the creeks rise and the ditches flood, long fingers of water reaching into the pinelands and quenching the thirst of the dry earth. During the months we'd spent in Costa Rica, wildfires had been burning in the mountains back home; the summer rains, come at last, soothed the embers and sent smoke drifting across the shoulders of the sacred mountains, looking like the mist of the cloud forest blowing up from Peñas Blancas. The world was wounded, but it was healing.

We adopted a puppy whose mother had died on the reservation, and we walked her in the prairie behind our apartment. We talked to the child that we'd lost, and I wondered if he could hear us; it felt like he was drifting further every day. On the wall in my office, I hung the topographic map of Monteverde and the Continental Divide, so that I could trace the backbone of the cordillera with my finger and remember walking in those enchanted forests, where we could almost hear the whispers of the dead. On the table near my desk, I set the photo slide that Marty Crump had given me, drawn out from her archives: dark forest, gleaming water, golden toad.

I had spent the last five years looking for the golden toads, but the golden toads were gone. Kyle had gone looking for their killer, but there wasn't a lone killer to convict; there was a disease that was made deadly by a changing climate, made global by a changing world. We had found no easy answers; we had witnessed no resurrection. But we had walked out of the jungles of the past alive and whole, which is more than most can say who have gone in search of El Dorado.

In all of the historical accounts and legends I could find, grief surrounded the golden city like a mist, and the explorers and conquistadors who went in search of gold and glory always paid a price. The English colonizer Walter Raleigh sacrificed his reputation and his fortune during his first expedition chasing rumors of El Dorado; as a consequence of his second expedition two decades later, he lost his son and then his own life to the golden visions that had haunted him like an infection in the blood. Another treasure hunter, the Spanish conquistador Antonio de Berrío, had already undertaken three expeditions in search of the city of gold by the time Raleigh found himself afflicted with the enfermedad doradista.* Berrío's obsession had been kindled by heredity and duty: After the death of his wife's uncle, Berrío had discovered a clause in the will compelling him to continue the late conquistador's search for El Dorado. Berrío died at seventy after two decades of searching, still mustering the resources for a final expedition. It seemed that the curse of El Dorado often punished the children for the sins of their fathers: Berrío's son died of the plague alone in an Algerian prison, captured by pirates during the course of his eighteenth expedition in search of the city of gold.

If the mythic city of El Dorado existed, holy grail, its trails were bewitched and its location enchanted, and the question that the explorer was meant to ask—*Whom would the grail serve?*—had always gone unspoken. And maybe that was always for the best. What would the explorers and conquistadors have done with it, if they had found it, after all? They could not have let it stand: El Dorado would have been

---

* Enfermedad doradista: Golden sickness.

plundered and broken, wergild for all the suffering and sacrifice it had cost to find it. They would have turned to rubble the one thing that they most desired. And what would come next? Would they be able to live with themselves, rich and powerful, burdened with the knowledge that they had brought ruin to the last great mystery? Gold and glory were compelling motivators, but some of those who had gone in search of El Dorado were drawn into the unmapped jungles by a softer call: a call of wonder, faith, and hope. It was something they could believe in, something they might have imagined in their childhood: a lost city, gold and glorious, waiting to be witnessed. But in the end they could not outrun the grief that the journey had awakened; it could smell them, like a dog in the night, and it always found what it was looking for.

In the weeks and months that followed our last quest to rediscover the golden toads of Monteverde, I could not help but feel as though I had also paid a price in my search for El Dorado. In the days that I had been out there in the forest, we had lost our unborn son: him drifting off to go back to whatever he had been before—a clutch of molecules—space-dust—afterlife. He was someone I would never meet now, and he would never hear the story; I would have to make my peace with that. And if the golden toads were really gone—as sad as it made me to believe that—then maybe it was time to let them go. Sometimes you lose something that you love, and you can't do anything to stop it. Sometimes you lose it before you ever really knew it, and it lives on only as a memory, or a sight and story passed down through generations, father to daughter, mother to son. And then some day even the stories end, and it can finally disappear—off to wherever the lost things go, out into the dark.

When we let go of the things that are gone, we can give our hearts to everything that's left. In Monteverde, they still find tapir tracks in the soft mud of the highlands from time to time. Jaguars roam the Peñas Blancas Valley like old kings below the forests that they once abandoned. The green-eyed frogs are coming back, and the brilliant forest frogs are climbing up the mountain on the Pacific slope; thirteen of the twenty-five lost frogs on Alan Pounds's survey list have reemerged. The resplendent quetzals are peering out at a changing world from their dark dens high above the earth, wondering if their wings will carry them above the peril. Karen Masters watches chanchos and pumas cross her camera traps on the trails above the station. The blue morpho butterflies that my father admired thirty-five years ago still fly around the fig trees at the foot of the Reserve. I have seen the place where a salamander lives in Bajo del Tigre, beside the ancient stone. Endemic orchids that you can't observe without a magnifying glass grow on fallen trees—life emerging out of afterlife. We could spend our whole lives looking for the golden toads, or we could take a walk in the cloud forest on a rainy afternoon to watch the sun set over the green hills of the eternal forest, loving it for everything it is, and not blaming it for what it isn't.

The truth is: They may be out there, and they may not; I can't say for sure. The forest that we explored on that forgotten mountain exists somewhere outside the boundaries of hope and grief: the last home of the golden toads should they ever manage to return. That forest is deep, and the cordillera is long. Perhaps they heard our voices from the foothills—the prophecies of doom—and chose not to reveal themselves; perhaps they left that mountain long ago, moving on to find new homes. We could choose to believe that, but we would never know.

As we had wandered down the mountain in the dark—our bootprints small in the feet of giants—I had asked Eladio if he would continue searching for the golden toads, when we had not found them at the fabled site and the world was not getting any bigger. "Possibly," he had answered, sounding tired but still hopeful. "That is the idea: to keep looking."

I have no doubt that he will, and if I hear word one day that he has found them again along the dark and mystic paths of the eternal forest, the news will not surprise me—but I do not know if it will comfort me. After all, there are other places, hidden places: unknown hollows, at the end of strange trails. And if it is not Eladio—if it is another treasure-hunter, seeking glory and resurrection from the lost wells of the cloud forest—I can only pray for caution, offering a whispered counsel: *Do not disturb the city of gold.* Maybe it is all best left to legend.

Before he had turned his expedition home, in the veil of rain and clouds, Walter Raleigh claimed that he could see the golden towers of El Dorado gleaming over Lake Parima. But the rivers rose with the holy rains, and he never reached the golden city that he could see beyond the hills. And if all we ever find is a distant glimpse of gold among the deep green forests—for a moment, while we are all alone, with no one watching—maybe it is better if we let that secret die with us.

As the summer ended and we watched the water quench the fires in the hills of cracked earth, Pri and I were not certain when our paths would take us back to Monteverde. We listened to the voice of thunder and watched the ditches swell with rain, and we wondered if the next rainy season in the tropics would send landslides running down the mountains to bury the old lost trails forever. We had hope that those forests would still be standing, whenever we found our way back into them—waiting for us, and for any others that might call them home.

Bijagua, Costa Rica

I want to tell you one more story. It takes place in the small town of Bijagua, not far outside Upala in the shadow of the Tenorio volcano in northwestern Costa Rica. The story begins in March 2018, when Donald Varela Soto heard a strange call rising from the water and reeds on an evening at the threshold of the rainy season. It had been a week of unusually heavy rains, and the frogs in the wetland sang in a wild chorus. Soto, out for a night walk to look over a sector of the wetland where he was planning to build a trail, knew most of those voices by heart: He had grown up in this countryside, and the trees, birds, and frogs were all familiar to him. All but one. Among the voices of the ten or eleven frog species that were calling from the reeds that evening, the new song was something he had never heard before. Curiosity got the better of him, and Soto waded into the shallow water to see if he might lay his eyes on the source of the strange call. He spent a few minutes reaching into the long grasses and stalking the boundary of the pool, but the voice had gone silent, and the frog evaded him. That was all right—he had other errands to attend to that night, and he thought he might come back the next evening to see if he could hear the song again. As he hiked back to the trail home in the dark, Soto was sure that he would have the answer to this riddle before long. It would take six months of searching to find the frog that had spoken to him.

Thirteen years before Soto first heard that new song drifting from the wetlands, the land that he was walking on had been a cattle pasture. The owner of the property had grown old and too tired to keep the cattle operation running, and none of his children was interested in carrying on the family farm; they had all moved on to the factories and rice fields in the lowlands years ago. When Soto visited the property that the old farmer had put up for sale, he and his friend identified more than 150 bird species in a single day; that alone was enough to convince him that the land should be protected. Soto and two colleagues pooled their savings and took out loans to purchase the property; and when one of the others began to move his own cattle in for farming, Soto offered to pay him a lease to keep the cows on another property: He was determined to devote the entire seventy hectares to conservation.

But Soto was not born a conservationist. "When I was growing up, I was really happy being a farmer," he remembers. "I grew up in a big family, nine to fifteen people in the house. This is how I grew up: I grew up farming the land."

It was his mother who encouraged his interest in nature and ecology, and when an opportunity arose for her son to take part in a three-month natural history course, she enlisted her friends in the Central Valley to help raise money for a loan to pay her son's tuition.

"That was the exact moment where I could change my life, and get more involved in conservation," Soto says. "Otherwise, I would have been just one more farmer, doing beautiful farming."

When you talk about Bijagua outside the community, Soto says, you usually have to get out a map and point out where it is. The region is primarily known for the Tenorio Volcano National Park, and the bright blue river cutting through it—Rio Celeste: the river from heaven. But the land that Soto and his colleagues had purchased, nestled in between the protected areas around the Tenorio and Miravalles volcanoes, sheltered other treasures. The forest that still stood in places around the cattle pastures was a biological corridor for the endangered Baird's tapir, moving between the slopes of the volcanoes. As the native plants and trees overtook the open fields, tapir sightings became more common, and interest from the local community around Bijagua grew.

"We invited people to come and see what we were doing," Soto says. "Anything we produce is shared with the community." The regenerating forests would draw visitors to the rural town—tourists who were looking for a more authentic experience, away from the crowds. In honor of the gardeners of the forest who had begun to reclaim the lands that they had once abandoned, Soto and his colleagues named the property the Tapir Valley Nature Reserve.

In the same way that the protection of the golden toad in Monteverde sheltered the native residents of those forests, the preservation of the Bijagua land for tapirs created a refuge for other species to survive or reemerge—some long forgotten, others never seen before.

One of those species was the frog that had been eluding Donald Varela Soto since he had first heard its strange call on a March evening on the borders of the wetland. All through the rainy season, he had been out in the night hours clearing trails around the property, and the song would often coax him in to the reeds and waters before going quiet just as he drew close; he was beginning to wonder if he was chasing a ghost. It was on an evening in October that he finally came face-to-face with the frog that he was searching for: He had followed the song like a spirit's call into the long green grass, and, after months of searching, the beam of his flashlight finally found its desire.

The small frog was bright green with dark spots speckling its back, a brownish streak running down from golden eyes. When it sang its song, its vocal sac swelled like a balloon, sending its strange messages out into the night. For the first time, Soto stood and listened to the call with the frog before him, and he was more certain than ever that this was a species he had never seen before. It would take more than three years of writing to biologists, sending photos, and monitoring metamorphosis, but in 2021 genetic testing confirmed what Soto had suspected from the start: This was a new species, never before known to science. Soto and his daughters would give the frog its name: *Tlalocohyla celeste*—the Tapir Valley tree frog.

Soto and a team of biologists conducted surveys in other parts of the forest, but they found no other populations; the species, it seems, exists only in that wetland: eight hectares in a protected forest, cradle for a new hope. Immediately following its discovery, the startlingly narrow range of the Tapir Valley tree frog meant that it was already critically in danger of extinction. Bijagua was not Monteverde, after all; its forests did not have the benefit of three decades of biologists advocating for their preservation. But to build a future for the bright green frog of Tapir Valley, the community of Monteverde offers one more lesson from the past.

In 1985, when the golden toads were still dazzling the highlands of the cordillera with their perennial glow, the Cloud Forest Preserve had been established to protect the rich forests above the small community, but the future of Monteverde was still far from certain. There were fires smoldering on the Pacific slope where the forests were being burned to make way for pastureland and agriculture, and on the Atlantic slope new settlers had begun cutting their way into the Peñas Blancas Valley. The biologist and photographer Patricia Fogden, who had been staying at Eladio's cabin in the valley to study sunbitterns, had reported that a homesteading family was busy taking over abandoned properties and cutting secondary forest to develop a community in wild Peñas Blancas; one night they had killed an umbrella bird and served it for dinner.

As the rainy season drew to a close, Alan Pounds had convened a group of locals* in the kitchen of the Pensión Quetzal in Monteverde to ruminate on these concerns. With the preservation of the Quaker Bosqueterno land and the creation of the Cloud Forest Preserve, it would have been tempting for the biologists and naturalists in Monteverde to sit back and bask in the glow of the protected forest: There was plenty set aside for scientific study, and enough biodiversity contained within the boundaries of the Reserve to keep a biologist busy for a lifetime. But that perspective had begun to ruffle the feathers of the community's more conservation-minded members. "Many of the biologists came, did their data collection, and left as quickly as they could," George Powell reflected later. "Any thought of the need to take care of the incredible beauty and complexity they were discovering did not manifest itself in any action. In fact, scientists were admonished in those days not to get involved in conservation because it would dampen their chances for tenure." There were others like Powell who had begun to feel a responsibility to protect and preserve their adopted home. What would be the future of the protected forest if it became an island? If the surrounding forests on the Pacific and Atlantic slopes were cut down for lumber and burned for agriculture, there would be no more biological corridors, no more genetic diversity, no more green mountain. Something would have to be done, they determined, and the small group of biologists, naturalists, and conservationists who had met at the Pensión Quetzal would not be enough to do it by themselves.

In the following months, the group continued to convene, adding more community members to their ranks; by the end of the year, they

---

* Among the biologists and nature lovers who had gathered at the little hostel were Bill and Barbara Haber, Willow Zuchowski, Richard LaVal, Bob Law, John Campbell, and Wolf Guindon.

had devised a purpose to their efforts: They would incorporate as a nonprofit organization with the goal of making more proactive land purchases for conservation in and around Monteverde. In February 1986, twenty-two charter members formally founded the Monteverde Conservation League.

By this time, the League had turned its attention largely to the Peñas Blancas Valley, where new reports of forest-clearing were making their way up and over the Continental Divide. A small contingent* from the Monteverde Conservation League descended into the jungles of Peñas Blancas to begin negotiating with homesteaders for the purchase of their claims. It was not an easy process; some landowners were willing but did not possess the legal papers for transactions; others resented the outsiders who wanted to buy them out of land that they had worked for months to clear. It is likely that the efforts of the Monteverde Conservation League in Peñas Blancas might have floundered in those early days, if it hadn't been for one local landowner and his little cabin in the valley.

Eladio Cruz had come over the Continental Divide into Peñas Blancas from San Luis with his father when he was twelve years old, carrying banana seeds. Before he met Wolf Guindon and began working with the Reserve, before he ever saw the golden toads, Eladio had been a homesteader in the Peñas Blancas Valley, carving a living out from the forests that would go on to shape his future. In 1986, Eladio became the first person to sell his property in Peñas Blancas to the Monteverde Conservation League; his old friend Wolf Guindon formalized the purchase. Between the two of them, the chain saw salesman and the homesteader planted the seeds of conservation that would grow to protect Monteverde over the next three decades. Following Eladio's example, dozens of other settlers in the valley lined up outside the Pensión Quetzal to sell their claims to the Monteverde Conservation League.

---

* Bob Law, Wolf Guindon, and Giovanni Bello; Bello later stated that they spoke with every farmer in the area, one by one.

But the original funds that the League had obtained from the World Wildlife Fund and the Audubon Society were running out; if the pace of land purchasing was going to continue, more resources would be required. Hope for the future arose from an unexpected visit. In 1987, a biologist named Sharon Kinsman was on a personal trip to Sweden when she was invited to give a slideshow presentation to local schoolchildren by Eha Kern, a teacher at the rural Fagerviks School. Kinsman had been living in Monteverde conducting research on cloud-forest plants, and her photos and stories from the dark and misty highlands of Costa Rica captured the imaginations of the Swedish students: They thought the forest even more precious because it was disappearing. When they learned that swaths of Monteverde's forest could be purchased for around twenty-five dollars a hectare, they started fundraising in their hometown, and Sharon Kinsman connected Eha Kern with the Monteverde Conservation League. The Swedish government agreed to match the funds raised by the students, and Eha Kern and her husband, Bernd, founded a nonprofit to organize donations; they named their organization "Barnens Regnskog"—the Children's Rainforest. Before the year was out, the Monteverde Conservation League was able to use the funds raised by the Swedish students to purchase six hectares of forest bordering protected areas in Monteverde.

The campaign begun by the Swedish students snowballed from there, gaining traction internationally; fundraising efforts spread to Scandinavia, England, Canada, the United States, Japan, Germany, and beyond. A Wisconsin teacher named Bruce Calhoun launched the "Save the Rainforest" initiative with his students. After a trip to Costa Rica with their children, Robin and Tina Jolliffe founded "Children's Tropical Forests UK"; when Tina passed away from cancer in 1992, their organization had raised $160,000. There was Chico Friends in Unity with Nature in California, Kinderregenwald in Germany, Nippon Kodomo no Jungle in Japan. In the end, these and other efforts raised close to two million dollars for the Monteverde Conservation League, enabling them to purchase 220 different properties—enough land to form a brand-new forest reserve in Monteverde.

In honor of the 554 hectares of watershed land that the Quakers had set aside for conservation in 1951, and in recognition of the dedication of the Swedish schoolchildren to protect a forest they might never see, the new reserve was named "Bosque Eterno de los Niños"—the Children's Eternal Rainforest. By 1998, the Children's Eternal Rainforest had grown to eighteen thousand hectares; it would swell to almost twenty-three thousand in the decades to come. The streams feed rivers on the Pacific and Atlantic slopes, running into seas on both sides of the continent. It contains the pond where Mark Wainwright rediscovered the green-eyed frogs, and the mountain where Eladio last saw the golden toads. It was—and remains today—the largest private forest reserve in Central America.

"Of course, conservation, in a way, only starts when you buy the land," Mark Wainwright will tell you if you slip in to one of his presentations for visiting students. "You can't just fence it off." Climate change, for instance, doesn't give much thought to the fences of a biological reserve; forests that have been protected from Homelite chain saws will still see their high pools evaporate in the dry years that are coming. If there is hope that those forests—and all the creatures that live within them—will remain, that hope hinges on the efforts of the people who know them best, and the support of others who have never seen them.

In the years following the creation of the Children's Eternal Rainforest, the Monteverde Conservation League embraced this philosophy. They worked with local teachers to develop a curriculum in the context of social needs and cultural values, leading school field trips to farms and forests and offering workshops for adults in the community as well. Collaboration with local farmers and landowners led to reforestation projects that planted more than a million trees and preserved corridors of fragmented forest between farms, along with a seedling nursery to safeguard the future of Monteverde's native trees. By combining international support with local buy-in, the Monteverde Conservation League has drawn a map to move beyond a fence-the-world-out approach to conservation, and the forests that it has protected are a living legacy for the local people and endemic species that call that mountain home.

That legacy has been carried on by others in the community. There is a kind woman called Doña Hermida who owns a polyculture farm on the Pacific slope of Monteverde, where she grows bananas, avocados, and chilies alongside the coffee crops among the trees; she doesn't use pesticides and she doesn't clear the forest to make room for her farm. There is a local climate-change organization called Corclima that is working to reduce greenhouse gas emissions in the community and cultivate support for climate action. The Mi Ocotea project is protecting and reforesting the endemic *Ocotea monteverdensis* trees in the cloud forest, which provide food for the resplendent quetzals. The Santa Elena Reserve—managed by the local high school—is protecting highland forest on Monteverde's Atlantic slope. And students from all over the world are traveling to Monteverde to participate in environmental education programs, sometimes staying longer than they'd planned, getting lost in the history and secrets of the cloud forests.

"It is so important that people understand that they CAN—no, MUST—make a difference," George Powell told me, fifty years after helping to establish the first forest reserve in Monteverde. "Neither Harriett, Wolf, nor I had any conservation training—we were regular people with a crazy idea. Anyone can do it if they want to."

So a group of locals met in someone's kitchen; a homesteader sold his father's land for conservation; and a biologist gave a slideshow to schoolchildren in Sweden. If any of these seemingly insignificant events had not occurred, Monteverde as we know it would not exist today. There might be an island of green forest at the top of a deserted mountain, black from fires and eroded from the feet of cattle; there might be the memory of a lost treasure, a eulogy for the departed species compelled to flee from ruin. Instead, there is an eternal forest, rolling like a wave from the Pacific slope up and over the Continental Divide and down into the Peñas Blancas Valley, washing up at the foot of the Arenal volcano. And Monteverde is not a last outpost of a disappearing kingdom, but a path that we might follow into better days ahead. Who knows where it could take us?

The story of Monteverde and the golden toad is often on Donald Varela Soto's mind as he works to ensure the survival of the Tapir Valley tree frog. While locally abundant—Soto reports more than fifty individuals sighted in a single night—until a second population is discovered, the Tapir Valley tree frog will remain one of the most endangered species on the planet. And while the Tapir Valley is protected, the outside world is closing in.

"In the last twenty years, the pineapple monoculture has moved in, and we have a lot of rice production in those areas," Soto says. "We've lost about thirty-five thousand hectares of wetland in the lowlands because of those monocultures. It kind of breaks my heart. We have no idea how many species have disappeared, because we had no idea they were there."

Tlaloc Conservation—an affiliate of the Costa Rica Wildlife Foundation—was created in the wake of the frog's discovery, and local Costa Rican biologists are turning their attention to the lowland marshes and wetlands that were once ignored and overlooked. And the local community around Bijagua has embraced the new icon of their forests. "The community realized they were helping to protect this new frog in the area," Soto says, "just because of trying to protect what they have."

Because his land is the only known home of the Tapir Valley tree frog, Soto feels a sense of responsibility and stewardship for the new species. "But there are some things that we can control, some things we can't control," he says. "We can't control what happens with the neighbors. One neighbor on the land above us is using selective herbicides, impacting the water. And right now we are having the driest year in the history of my town. We don't know how the species will react to this."

When I spoke with Soto over the phone in May 2024, he had recently returned from filling reservoirs in the wetland by hand, adding water to keep the pools and marshes full. When we finished our conversation, he would fill a few more buckets and walk back out into the forest, bearing a drink for a wounded land.

When I asked Soto what he would want one to learn from the story of the Tapir Valley tree frog, he answered: "Your individual actions count. You have to do your part. And I think this is what I have been doing: I have to contribute to my community, I have to contribute to my family, I have to contribute to conservation. It's something that was passed on from my mom and my granddad, and I am trying to pass it on to my daughters."

When Soto discovered the new species, his daughters were seven and nine. They followed his footsteps through the long grass to dip their hands into the water and draw out tadpoles, which they watched transform in metamorphosis, new generations, lineage, heredity.

"Now they are always asking about conservation and wildlife," Soto says. "And maybe this could inspire others, people in rural areas—maybe they are not biologists, they are not herpetologists, but they are always in the field—maybe this could inspire them to keep looking and keep doing what they're doing. One day you might be able to protect it."

Donald Varela Soto grew up in Bijagua, and Bijagua was home to the forest that he would go on to protect. But as a young man, he had worked as a nature guide in another community: Monteverde. And when he discovered the new species in the wetland of Tapir Valley—a species that had eluded even one encounter with biologists in its entire history—he couldn't help but think about the golden toads. If a brand-new species could be discovered in the countryside, in a wetland that had been a watering hole for cattle just a generation earlier, maybe there was hope again for a little golden toad that had disappeared in the cloud forests of Monteverde, thirty years before.

Monteverde is a place where things disappear, and a place where things persist. There are frogs in those forests that have lived far longer than we have, and maybe species we will never know. There are big trees more ancient than the oldest hunting trails, thick roots reaching deep into the earth.

Eladio's little cabin is still down there in the Peñas Blancas Valley, succumbing to the elements the last few years—the floorboards are growing soft, and the roof is letting water in like teardrops falling on the wood. It has been ten years since I have been there; I wonder if it will be there ten years from now. It is a fragment of his childhood—and of his father—that Eladio has given up so that it might be part of something greater, something that will, with any luck, outlive him, and me, and anyone who comes there after us. And if the forest overtakes it—creeping vines and roots and branches burying the dead—it will be a fragment shored against its ruins, a collapsing temple in the shadow of an eternal forest.

Walhalla, South Carolina

A small car drove west through the Carolina foothills, following the old familiar roads into the mountains. In the early morning, the summer sun flickered through the canopy of green trees that hung above the country road, and the branches were like arms reaching out for one another, scattered relatives leaning toward reunion.

Kyle had been driving since leaving our parents' house in the upstate, and our father was in the passenger seat, craning his neck from time to time to look for circling hawks riding the thermals coming off the highway's heat, or the glint of red from a summer tanager among the leaves. Soon they would cross Lake Hartwell, where our father used to take us fishing on the red-clay shores. They would pass near Saddler's

Creek, where we had ridden bikes together on the forest trails. They would cut through Townville, with its fields of tall grass and long canals, where our father had once chased his own holy grail: a hybrid butterfly, part viceroy and part red-spotted purple, that had been rumored to be seen from time to time above the marshes on the outskirts of the town. Kyle and I had been nine or ten years old then, and we spent the day seeking shade and shelter in the dry land as our father chased the wild myth with a butterfly net up and down the outskirts of the forest; in the end, he had given up the quest to take us home for dinner. We had not been able to see what he had seen, back then, and all that he was trying to reveal to us.

Now three years had passed since Kyle and I had walked down from the mountain of the golden toad together, our search accomplished and our hands empty. I had gone back to my home in Arizona, and Kyle had traveled east to the Atlantic coast, where he and his wife rented a small house between the forest and the sea. In the summer, he followed the coast south to the Carolinas to visit our parents and the town that we'd grown up in, and to go looking for something that had not yet been lost.

They turned north onto Highway 11, and the woodlands fell away to open fields of churned earth, like dug-up graves. A sign on the fence-line read, *This ideally situated estate comprises ten acres of excellent building land.*

But ahead of them they could see the Blue Ridge Mountains, hazy in the distance. Their destination lay in a national forest just past Walhalla near the North Carolina border, and any map printed in the last two decades would call it Station Cove, but our family knew it by another name: Salamander Falls.

We didn't understand it at the time, but Kyle and I had grown up in opulence. When we were young and we wandered into the mountain streams to flip a log or look under a rock, we didn't think it strange at all to find a small, dark salamander looking back at us. That's because the southeastern United States is the number one hot spot for salamander biodiversity on the planet. If you were to look at a heat map of the globe, you'd see a world of cool blues and muted grays, broken by two splashes of color: a brushstroke of bright yellow smeared along the East Coast of North America, and at its center, in the heart of the Appalachians, a warm and flaming ember of deep orange, smoldering like a fire. There are more species of salamanders there than any other place on Earth; nowhere else in the world comes close.

Out of the 550 salamander species known to science worldwide, more than a hundred of those are native to the southeastern United States. And out of that hundred, many species are endemic to the streams and rivers of the Southeast; they are like the golden toads in Monteverde, believed to exist in just one place, a singularity. The Pigeon Mountain salamander is known to just a single mountaintop in the state of Georgia. The Black Warrior waterdog is only found in the Black Warrior River Basin in Alabama. The South Mountain gray-cheeked salamander; the Peaks of Otter salamander; the Caddo Mountain salamander—the list goes on.

Growing up in the foothills of those rich mountains, Kyle and I had lived with salamanders at our fingertips—they dwelt in the caves and hollows a short hike from our back door. And even more importantly, we'd grown up with parents who could reveal them to us—a mother who would lift us up to peer into the dens of mystery, a father who knew the name of every creature and spoke of them like old friends. It's a kind of splendor that you don't appreciate until it's gone.

It was beginning to look like rain by the time Kyle and our father made it to the trailhead at Salamander Falls. They started down the trail as bulkhead clouds overtook the sun, and they followed the winding path into the woods until the sound of the highway was lost behind them, the rising and falling whistle of a hermit thrush (so like the squeaky wells of the black-faced solitaire) following them along the trail. From time to time, our father stopped to overturn a log, or bent to peer into the dark cavern underneath a rock—looking for skinks or copperheads. He had always done this, ever since Kyle and I were young; in the woods, he was less interested in our destination than what he might find along the way. He had taught us the names of the bright butterflies and the old snakes that lay sleeping in dark places. At the bottom of the waterfall where they were heading now, in the mossy dens where the water crashed to earth and sent mist drifting in the air, he had shown us the little creatures that made their homes in the wet rock, delicate and beautiful—when we were young, we'd called them "salamamas." His first love would always be butterflies and moths—those temporary angels that had kept him alive through the dark winters of his childhood—but to us he was the Salamander King.

They crossed the old boardwalk bending low above the creek bed; in some places, the lumber had decayed, and there were pieces missing. Kyle watched our father touch the handrail absently, his other hand at the small of his back; he had hurt it years ago and it had never really healed. It had changed the way he walked, and he wasn't able to bend low to look beneath the leaf litter for black widows or trapdoor spiders like he'd used to. It was something that continued to surprise us, whenever we came home to visit: Our parents were getting older. When our mother had visited Monteverde a few years ago—thirty-five years after she'd first seen the green mountain—it had taken her a long time to walk up to the ventana along the old camino—she who used to carry the two of us together in her arms—her steps more careful and her breathing hard. And watching our father and his careful steps, Kyle couldn't help but wonder if he might have climbed into the highlands of the golden toads with us after all, if only we'd invited him. What magic might he have carried with him if we'd been all together then—old magic, family magic? What might he have seen among the leaves and pools that we had missed?

Beneath high, green leaves, Kyle and our father crossed the final stream over slick rocks and went on up the narrow trail, toward the mounting roar of falling water, and stepped out at last into the view of the falls. The churning white water came down over tiers of dark black shining rock; here and there, deep caverns scarred the wall, curtained by the spray of falling water. You could imagine dark eyes looking out from those caves, even if you couldn't see them. When we were young, we'd imagined that old monsters made their homes in those shadows; now, of course, we knew that the monsters were the kind we couldn't see.

Kyle and our father stood alone before the calm pool of swirling water that collected downstream of the falls. Here and there, our father knelt down and squinted into the cracks and crevices of piled rocks and fallen logs, searching for a flash of glistening skin or the flicker of a salamander moving in the dark. Uncertain if he would find what he was seeking, Kyle climbed up to put a hand on the damp green moss growing on the black rock, water falling all around him like a downpour. He bent down to peer into a likely hollow: nothing.

The rock in front of him grew darker, and he felt the temperature drop as the gray clouds drifted across the sun. A looming thunderstorm preparing to break, to flood the land, to wash it all away. It was like the specter of *Batrachochytrium salamandrivorans*, across the sea—inevitable, unstoppable.

"When it gets here," Joseph Mendelson had told Kyle, "hell's gonna break loose."

"We've been operating with the understanding that it's just a matter of time," Mark Mandica had told him. Once the steward of Toughie, the last Rabbs' fringe-limbed tree frog, Mandica is now the executive director of the Amphibian Foundation in Atlanta. "*BSal* will get here eventually."

Today, still carrying the weight of Toughie's loss, Mandica runs educational programs for young herpetologists and supervises conservation projects for threatened amphibians. One of the species that he manages—the frosted flatwoods salamander (*Ambystoma cingulatum*)—has suffered a 90 percent loss since the year 1999; but in 2022, it bred in captivity for the first time. Successes like this give hope that, if and when *BSal* does make landfall in the Appalachians, we will not be taken wholly unaware. That is the dark gift that *Bd* left for us.

Mandica is not the only one working to stave off an invasion. Back in 2018, when Pri and I were searching for maps of the Brillante highlands in San José, Kyle met JJ Apodaca at a coffee shop and brewery on the shores of North Carolina's French Broad River. Apodaca is the executive director of the Amphibian and Reptile Conservancy, an organization implementing local conservation projects in priority amphibian and reptile conservation areas. Recently, they have been working to reintroduce the frosted flatwoods salamander in the wild; it is a species that seems to glimmer with veins of diamond, and it hasn't been seen in the wild outside Florida for ten years or more. They have also launched projects in North Carolina's Hickory Nut Gorge. A twenty-thousand-acre, fourteen-mile-long canyon in the Blue Ridge Mountains, the gorge is home to the Hickory Nut Gorge green salamander, the crevice salamander, and the Blue Ridge gray cheeked salamander, and a staggering diversity of regional species.

The best thing we can do to prepare for *BSal*'s arrival, Apodaca told Kyle, is to strengthen southeastern salamander populations, so they are able to fight back against the pathogen themselves.

"There's one way to really prepare," said Apodaca, "and that's to leave large populations with lots of genetic diversity."

A key factor in that campaign is the management of protected—and connected—habitats, and the Amphibian and Reptile Conservancy is developing new ways to monitor those keystone forests.

"It all comes down to the choices we make about protecting the integrity of the habitat that these species rely on," says Ben Prater, the Director of Southeast Programs for the conservation organization Defenders of Wildlife.

Since the discovery of *BSal*, Defenders of Wildlife—in partnership with biologists like Karen Lips and George Rabb—has been advocating for better regulations in the pet trade, hoping to mitigate the risk from infected amphibians carrying *BSal* into the United States. They achieved a partial ban on the import of salamander species in 2016, but the organization considered this to be only the beginning of their efforts; more important are the local, on-the-ground efforts that they undertake to ensure healthy and intact salamander populations in the wild.

"We're gonna keep fighting to make sure that the salamanders and the unique habitats they need have a voice as well," says Prater. "I believe in the adaptive and the resilient aspect of nature, and have always put a lot of credence in the belief that if we protect enough habitat and ensure the quality of those habitats, then there will always be a future for salamanders."

"The southeastern U.S. is probably where salamanders speciated," says Will Harlan, Southeast Director for the Center for Biological Diversity. "They really came from here, and this is one of their greatest strongholds. So it's crucial that we protect them and their habitats here."

Recently, the Center for Biological Diversity petitioned to protect two salamander species under the Endangered Species Act: the Hickory Nut Gorge green salamander, and the yellow-spotted woodland salamander. The latter was only scientifically described in 2019, and recognized right away as a species in need of conservation. In addition to these protections, the Center for Biological Diversity is also advocating for the protection of national old-growth forests.

"Salamanders need mature and old-growth forests, especially here in southern Appalachia," says Harlan, "with the headwater streams, the seeps, the springs. That's prime salamander habitat, and you're not gonna find that anywhere else. So that's why the Center for Biological Diversity—along with the Southern Environmental Law Center, Sierra Club, MountainTrue, and Defenders of Wildlife—have all joined together to force the forest service to create a better plan that protects more of the forest. We're fighting to ensure that those key places the salamanders depend on are protected."

THE ETERNAL FOREST

Some of the people fighting to protect the salamanders wear expensive suits and lobby for legislation in the courts; some peer through microscopes to develop vaccines against the pathogen; some dedicate their lives to the captive breeding of endangered species; some experiment with practices like cloning and resurrection* that lie at the edge of wild science; and others work to ensure that these amphibians will have a place to call home, for as long as they are able.

In an op-ed in the *New York Times,* Joseph Mendelson and Karen Lips offered a warning and a plea in the shadow of *Batrachochytrium salamandrivorans,* reflecting on a past they cannot change:

> *That invasion continues to haunt field biologists. We encountered tropical forests practically devoid of amphibians that were once teeming with dozens of species of them. We watched mass die-offs. We tried to save threatened species by airlifting them out of infected areas, breeding them in captivity and searching for answers through field and lab research.*
>
> *None of it worked. No cure exists for wild populations. Losses of amphibians continued around the globe. We've seen no significant recovery of populations. Worse, the fungus persists in the environment, preventing the reintroduction of captive animals.*
>
> *This time, we have ample warning to prevent the arrival of Batrachochytrium salamandrivorans into the United States. We know what kind of killer we're dealing with. A global network of biologists is studying its movements. Government agencies are on the alert.*
>
> *Let's get it right this time.*

. . .

---

* At the University of New South Wales, Australia's Michael Archer has been working for over ten years to create a living specimen of the extinct southern gastric brooding frog from frozen DNA. "The work continues on the Lazarus Project," he reported in a personal communication in 2024. "It is currently being led by a Ph.D. student. . . . This student has just achieved a breakthrough that is highly encouraging."

It was early afternoon when the storm broke over Salamander Falls. Kyle had almost given up hope of finding anything, but our father was still moving in a hunched form among the waters, peering into the hidden dens—still searching.

Twenty years ago, he used to take us out to the gas station down the road from our house in the early mornings because its bright lights would call in moths from the heavens, dozens of them coloring the white walls. Sometime in the last two decades, the gas station had started spraying pesticides, and the moths had disappeared. But every time he passes the building now (or so my mother tells me), he still wanders over to search the boundaries for signs that his old friends have returned. Someday, if the salamanders vanish in the dark, I imagine that his faith will still draw him up to Salamander Falls to try to find them. He is the kind of father who will never stop looking.

The thunder spoke from high above the steep walls of Salamander Falls, and as the first drops of rain began to fall, our father waved to Kyle from the riverbank. Kyle scuttled down the steep rocks and balanced over the wet log growing mushrooms; stone by stone, he crossed the water to our father. He had raised his hand and was pointing to a crevice in the rocks, where the rain had drawn out from the shadows a small salamander—the first of many that would emerge to greet the rains. He was looking out at the world with eyes of hope. It was a beautiful day.

They stood there together at the bottom of the waterfall, listening to what the thunder said, and a summer storm came down from the far blue hills to touch the earth.

EPILOGUE
EL DORADO Y LOS ÁNGELES

Los Angeles, California

I am going to see the golden toads. The thought sent a feeling of cold water running down my back. It was a truth that I had only ever felt in riddles, in wonderings and tempered faith: *If I climb the right mountain at the right time, will I find them?* But here at last, I was certain: Today, I would see the golden toads. How long the path had been that led me here.

I was closer to the lost species than I had ever been before, but I was a long way from the cloud forests of Monteverde. I was sitting in the rose garden outside the Natural History Museum of Los Angeles County, in the shadow of a eucalyptus tree and at the foot of a building that looked like a mausoleum. In a few minutes, I would go inside and look into a jar, and the long path would come to its end.

I had driven out from Northern Arizona, west on I-10 through the southern part of Joshua Tree, and crossed the Mojave Desert as the setting sun sent orange and purple colors out across the big, empty sky like the world was on fire. They were the same long, straight roads across the desert that Pri and I had driven back in 2019 to meet Jay Savage at his home in San Diego, where old shrines to the golden toad sat like funeral

EPILOGUE

pyres among a maze of filing cabinets that had swallowed secrets we would never know—a whole life's worth of research, memories, and lore.

This time I'd come out alone. I'd slept in the desert the night before, among the bones of the earth, trying to pray. Pri was home in Flagstaff; Kyle was on the East Coast, looking at a different ocean; I would have to walk the last few miles of the journey on my own.

I walked up below the steps of the Natural History Museum, watching families come and go, and the vendors selling fruit and ice cream from their little carts. A little after ten, I wandered into the rotunda, feeling small beneath the dome of eternity. Neftali Camacho, the Herpetology collections manager at the museum, came and found me after a few minutes; he led me through the rotunda where the skeletons of dinosaurs looked on from their long halls, towering over a lost kingdom, old grief preserved in stone. We passed through a door that led to dark, narrow passages dropping down in catacombs of stairs, and the distant clamor was lost behind us. The twisting corridors reminded me of the old trails: Brillante, Chomogo, Rincon. But there was no bright mist here, no water in the air sweeping up from the distant valley. These were the halls of history, and they led to El Dorado.

Neftali stopped in front of a small door, opened it with a key, and stepped inside. I followed him.

There were crocodile skulls on filing cabinets and the shells of leatherback turtles on the wall; and down the hall, the tall sliding cabinets of infinitude, full of secrets. Neftali led me to a shelf with "Salientia" written on the cabinet—"Bufonidae, Centrolenidae, Dendrobatidae..."—and reached into the cavernous dark to draw out three jars—one large, one medium, one small. Then he turned and carried them into the light. There, on the silver table in front of us, he set the golden toads.

They were floating weightlessly in glass jars of ethanol, lost in space. In the biggest jar, there were maybe a hundred males, packed together clambering over one another as if they were trying to escape. Their magnificence had faded a little over the years—bright orange tarnishing to

EPILOGUE

yellow—but they were still as beautiful as I'd imagined them. They were smaller than I'd thought they'd be: more vulnerable, more fragile. They were tangled in strings like puppets. Some of them had their mouths open, as if they had died speaking; now that I had finally found them, I couldn't think of anything to say.

The second jar was full of females, maybe twenty-five in all. Their dark eyes were clouded from the alcohol, or from having looked beyond death. The third jar contained a handful of juveniles. I had heard Eladio talk about these, but I had never seen them, not even in pictures. They were a dull rusty color, spectacled on the belly, and very small—probably about the size our son had been when his heart stopped beating.

After a few minutes of searching, squinting into the distorted perdition of the jar of golden males, I found him.

The small tag on his leg read "1893." He was the first specimen collected by Jay Savage, Norm Scott, and Jerry James—May 14, 1964. When Savage had left Costa Rica to return to the United States, he had carried this toad with him to show the world the treasure that was buried in the cloud forests of Monteverde: sign of glory, holy grail. Now it had gone on into whatever afterlife awaits the toads, and it had left its body here for me to find, here among the memories preserved in jars. It would stay here forever; it would never find its way back home.

On that first trip, Savage and Scott had collected 142 male toads and 31 females, and most of them—those that had not gone off to the distant great museums of the world—were here.

All that the golden toads had ever heard and seen and felt was lost now to the ages. Their time had ended, and these were the first of them and the last of them. But I remembered everything—every story that I'd ever heard from the biologists and campesinos, storytellers, tour guides, children, mothers, fathers. The golden toads were gone, but their stories would survive—emerging now and then from dark hollows when the patter of rain began again on green leaves in the mist, emerging from a den of secrets, bringing light back to the forest.

EPILOGUE

I stayed there in the collections room for a little while, looking at the golden toads. I knew it might be the last time that I ever saw them, and I wanted to remember it forever. They didn't feel particularly mystic, didn't bring enlightenment or happiness or peace. They were just toads, in the end—once golden. Why should they be any more than that? I tried to memorize their images, so that I might tell my father all about them, the way that he had once told me. And then, having heard all that they had to say, I walked back to the hallway, and Neftali returned the jars of golden toads to their resting place among the others in the cabinet. The door closed behind us, and I did not look back.

. . .

That afternoon, I walked out to the pier on the Pacific Ocean to watch the fishermen cast their long lines in the waters. The August air was light and warm, and a cool breeze with a taste of salt came off the ocean. The sight of the golden toads had sent my thoughts back to the cloud forest, and I was thinking about something that Eladio had said. On our long walk down from the forgotten mountain, I had asked him what he would do if he ever found the golden toads again, his old friends who had left him long before. He had thought about the question for a long time before he answered.

"If I found it within the borders of the Children's Eternal Rainforest," he had told me, "where I knew it could be protected, I would tell the world." Then he had fallen silent, looking with a smile upon a vision only he could see: a flash of gold in a dark forest, leading back to splendor. And then he'd added, "If I found it somewhere else, I would stay quiet."

Below the cry of gulls, above the waters, I walked along the boardwalk back to shore. The old fishers cast their lines, and the touch of the sun healed old wounds, in a dry land that lay before the sea.

For three years, I had been thinking about the afterlife: where we might go when everything ends. It was strange to think that there were

EPILOGUE

hearts out there that knew the answer. The golden toads knew. They had learned that great secret when they descended for the last time into their deep dens beneath the earth, hope being the last thing that they lost, leaving everything they'd ever loved behind.

In the ages when the elfin forest was young and wild and silent, save for the songs of the first birds and the drum of rain and the howling of the immortal wind, they had been a gold light glimmering low and deep in the dim green of the mountains. They were the first fires that lit an endless night, signal pyres that had come out from under rock and root to feel the breath of rain against their skin and see the starlight in the heavens. They tucked their beloved eggs in pools of mystery and conjured rituals. They gave light to a dim world. Their fire spread, dispersed, diminished. They left that forest, and their spirits drifted off. They went beyond that forest; beyond the clutching roots and the planets of dew—shining spheres reflecting galaxies. They were weightless. They were unassailable. They went beyond grief, beyond despair, beyond extinction and eternity. They became memory, mythology, heredity, existence. They became the spirits of the celestial mountain, the mystic sentinels of a high enchantment.

I dream that the golden toads move in tunnels that run through time in the cloud forests of the afterlife. Dirt walls, warm and winding, ceilings like a low cathedral—they are scattered with the footsteps of the ones who have gone before. Those tunnels weave in tangles, deeper, farther, until their long fingertips grow cold, touching the ancient waters. Concealed in the shadows beneath green leaves, in the wavering pools, the wind breathes life into a lost species. Their ghosts come down to touch the mountaintop, carrying word from the beyond into the corporeal forests, prophecies to those that still sleep beneath the stars. They are the golden shapes of the sunlight settling through the canopy in the charmed air of a tropical evening; they are fire sermons drifting in the cold, high dark.

For all the time we knew each other, our two species were always sundered. We never learned to tell them how magnificent they were

EPILOGUE

to us, in a way that they could understand. They never learned how to call for help. We only knew one another from afar, all that we could learn from our studies and explorations. We might wonder if we could have tried harder to understand them—or would they be unknowable, forever beyond us, vanished and eternal? When everything ends, and if we find each other again, maybe we will remember one another. Maybe we will reach across the gulf of speciation and share a fire. It will only be a trick of the light, but it will look like a lonely figure bending down to speak with a golden toad, two phantoms spotted for a moment, then lost again in the flickering evening light.

Life, death, afterlife. Egg, tadpole, toad. Orange sky, green leaves, and shadows—in the half-light, the shyness of the branches makes the forms of golden toads.

They might be only space-dust now, or they might be waiting until the whole world just goes still; waiting until the roll of thunder is the only voice that echoes in the highlands; waiting for the moving stars to go out one by one, for the cries of sorrow to fall silent. A great sigh, like a wind, will come through shaking the grass, and beads of water will fall on the dry earth.

Then, out of their distant tunnels, the golden toads might emerge again—long after we have gone away to other lands, or other worlds, or our oblivion—and look out from the cusp of a wild forest to a familiar valley; they might feel the fog and mist blowing up from the windward slope through the cool veil of an eternal forest: and at last a damp gust, bringing rain.

. . .

Now it is close to midnight, and the stars are out. I have just come back from walking in the desert, northern Joshua Tree, on my way home from California. The night wind is shaking my tent, sounding like a banshee in the dark. Ten thousand miles away, the first frogs are coming

EPILOGUE

out to greet the evening rains. But I have seen the golden toads at last, and this is where the story ends for me.

It is three years since I was last in Costa Rica; I have not returned since our expedition with Eladio to his lonely mountain, the same year that we lost our son. I did mean to go back. I often took the old maps out from the drawers where I had hidden them, to touch the invisible trails that I had hoped to walk again. I had wanted to go to Quepos to look for the variable harlequin frogs, and to visit the Tapir Valley Reserve and see the new tree frog for myself. I wanted to page through the lost papers in the library at the Reserve in Monteverde—to read what the first biologists had written of the ridgeline more than fifty years before. I wanted to walk with Eladio again in the old familiar forests, and ask him what stories he still knew of ghosts and grief and making peace with the dead.

I had spoken all of this to Pri, the deep desires of which I was ashamed, and she had never asked me not to go; but she had told me: "We've given six years of our lives to the golden toad. That's enough."

So, instead of going again into the tropics, I will drive home in the morning, and in the evening I will walk our dog out to the cattle tanks in the fields behind our house; we will look for tiger salamanders in a few inches of water, and listen to the chorus frogs as the sun goes down beyond the trees. I am finished searching; I have found my answers; it is not my story. I have looked into the lost reflecting pools and I have asked the question—*Whom does the grail serve?*—and the answer has come back to me from farther forests in a language I can never translate: a distant trill and a soft cry, *tep-tep-tep*, wooden spoons tapping in the dark. I am going home to set my lands in order.

We got the news in December, you see, and in four weeks Pri will give birth to our daughter, hallelujah, shantih shantih shantih, gracias a Dios. And the next time that I go back to Monteverde, I will not be alone—I will show her the cloud forest and the old trails, and she will see the mist coming over the Continental Divide from wild Peñas

EPILOGUE

Blancas into the eternal forest. And when she is old enough to understand it all, I will tell her all the stories that I've ever heard, and show her the images of splendor, and sing the songs—the way my father told the stories to me and Kyle after we had come into the world together: two kids watching the dust rise in a pillar of light from a projector, unveiling the world's lost gold. I will tell her everything: the cold forests at the top of the world, resurrection, grief, fantasmas, el dorado y los ángeles.

And the golden toads will, in a way, return.

ACKNOWLEDGMENTS

We went into the jungle (and the desert) when we were twenty-two, and when we walked out, by God, we were rich. The best of those riches have been the family that we found there. To Pri, without whom this book would not exist, thank you for your support through a decade of golden toads, and for bringing ALRP into the world: the best treasure we could hope for, better than gold. To Alannah, thank you for your healing magic, and for always seeing what others couldn't.

We may not have gone into THE TROPICS at all if it hadn't been for the invitation of Karen and Alan Masters; thank you for lighting the path into the jungle and offering a truly life-changing experience, and thank you for the advice, encouragement, and words of warning on the trail of el dorado.

Thank you to everyone who welcomed us into Costa Rica (Moncho, Gisella, Raquel, Johel, Adam, and Pao) and to the friends who made Monteverde feel like home (Juli, Hazel, Nicole, Ana, Vanessa, Hector and Evelyn, Marcela, Pam, Alexa, and all the others). Even a summary of our gratitude would be too long to print.

Thanks to the friends who read the early drafts, talked through big ideas, and swung their machetes to find the way through the bad paths (Coleman, Margaret, Megan, Emily, John, Peter, Bryan, Steve, and Shanon), and to the SC contingent for years of support.

At UCR, thank you to Federico Bolaños and Gerardo Chaves.

ACKNOWLEDGMENTS

For making connections and providing archival records, thank you to Bronwyn Mitchell and Janelle Devery.

Thanks to the Children's Eternal Rainforest and the Monteverde Conservation League (to Lindsay Stallcup in particular) for supporting a hike into a protected forest, and for keeping it eternal. Special thanks to Mark Wainwright, Luis Solano, and Gilbert Alvarado for lending your expertise to the search for the golden toad.

To everyone who lent their voices to this tale—Marty Crump, Jay Savage, Norm Scott, Frank Hensley, Alan Pounds, Ricky Guindon, Hazel Guindon, Giovanni Bello, Marvin Rockwell, Katy VanDusen, Robin Moore, George Powell, Donald Varela Soto, Mills Tandy, Justin Yeager, Cesár Barrio-Amorós, Mynor Kasho, Ulises Morales, Greg Czechura, Allan Pessier, Joyce Longcore, Lee Berger, Joseph Mendelson, Forrest Brem, Dede Olson, An Martel, Mike Sears, Ben Prater, Will Harlan, JJ Apodaca, Mark Mandica—thank you for trusting us to share your stories. A book could be written for every one of your tales.

Thanks to Robin Moore and the Re:wild team (especially Lindsey and Devin) for supporting the search for the golden toad as part of Re:wild's Search for Lost Species, and thanks to Esteban Brenes-Mora for saying, "There's funding for that."

For the excellent reporting that eased our research, thanks to Kay Chornook and the Guindon family for *Walking with Wolf*, to Nalini Nadkarni and Nathaniel T. Wheelwright for *Monteverde: Conservation of a Tropical Cloud Forest*, and to the editors of the *Monteverde Jubilee Family Album*.

At Northern Arizona University, thanks to Mark Neumann, Kurt Lancaster, Peter Friederici, Annette McGivney, Nicole Walker, Geetha Iyer, and KT Thompson. A "Support for Graduate Students" grant enabled the first trips back to Monteverde.

Thanks to Alexa Woodward, Kolea Coody, and Eric Lynch for sharing your music.

ACKNOWLEDGMENTS

Thanks to Sumanth Prabhaker at *Orion* and to Ash Davidson for encouraging the story.

At Diversion, thanks to Liz, Evan, Nina, Scott, and Keith. At Neuwirth & Associates, thanks to Amy, Beth, and Jeff.

Extra special thanks to Lauren Hall at Folio, for seeing what we saw.

Most importantly, thank you to DBR and RLR for telling us about the golden toads and showing us the salamanders; everything we are, we owe to you.

And finally, thank you to Eladio, for sharing your story.

This is the story of the golden toad and the people who knew it best; any of its triumphs are theirs, and its shortcomings are our own.

ENDNOTES

ARCADIA

viii **"In the same epoch . . ."** General evolutionary timeline for Bufo periglenes is derived from a personal conversation with Joseph Mendelson via email in 2019 and this article: Ines Van Bocxlaer, Simon P. Loader, Kim Roelants, S. D. Biju, Michele Menegon, and Franky Bossuyt, "Gradual Adaptationwhat toward a Range-Expansion Phenotype Initiated the Global Radiation of Toads," *Science* 327, no. 5966 (February 5, 2010): 679–82, https://doi.org/10.1126/science.1181707.

PROLOGUE: AN ORANGE BUFO FOREST

Unless otherwise noted, details in this chapter were derived from personal communication with the following sources:

    Neftali Camacho, 2021 and 2024   Norman Scott, 2021
    Jay Savage, 2019

xi **"the runway of the new . . ."** Alejandro Zúñiga, "Costa Rica's La Sabana: From Airport to Park," The Tico Times, October 28, 2019, https://ticotimes.net/2019/10/28/how-costa-ricas-first-international-airport-became-its-biggest-urban-park. xi **"above the Pass of Desengaño . . ."** Jay M. Savage, "Preface," in *The Amphibians and Reptiles of Costa Rica: A Herpetofauna between Two Continents, between Two Seas* (The University of Chicago Press, 2002), xi. xi **"That year, Irazú was erupting . . ."** K. J. Murata, C. Dondoli, R. Saenz, "The 1963–65 Eruption of Irazú volcano, Costa Rica (the Period of March 1963 to October 1964)." *Bulletin of Volcanology* 29, no. 763 (1966). https://doi.org/10.1007/BF02597194. xi **"Savage had been skeptical . . ."** Jay M. Savage, "An Extraordinary New Toad (*Bufo*) from Costa Rica," *Revista de Biologia Tropical* 14 no. 2 (1966): 153–67. xi **"generally unremarkable in appearance . . ."** In general, earth-tones allow toads to blend in with their surroundings, and most toads in the Americas exist within a more limited color spectrum of gray, green, and brown. Savage wrote in his paper describing the new species: "Toads of the genus *Bufo* are not noted for their brilliant coloration nor marked sexual dichromism. Most members of the genus are dull creatures essentially gray or brown in ground color with dark blotches and sometimes light markings. Within the group a number of species show slight sexual dichromism, but relatively few approach the condition in such forms as the Asian *Bufo raddei* Strauch or *Bufo canorus* Camp of the high Sierra Nevada of California. Even in these latter species the colors are subdued greens, olives, and dark to rusty browns. In striking contrast to the usual *Bufo* situation, the new Costa Rican toad described below exhibits the most startling coloration and development of sexual dichromism recorded in the genus." Ibid., 153. **xii "collectors had traveled relatively deep . . ."** Amelia S. Calvert and Philip P. Calvert, *A Year of Costa Rican Natural History* (New York: The MacMillan Company, 1917). Daniel H. Janzen, *Costa Rican Natural History* (The University of Chicago Press, 1983). **xiv "They left the city at one . . ."** Charles McLaughlin and Fred Truxal from the Los Angeles County Museum joined Jay Savage and Norman Scott on the trip to Monteverde. Savage, "An Extraordinary New Toad (*Bufo*) from Costa Rica," op. cit., 153–67. **xiv "a government tractor had worked . . ."** Lucky Guindon, "From Old Letters by Lucky Guindon," in Monteverde Jubilee Family Album, edited by Lucille Guindon, Martha Moss, Marvin Rockwell, John Trostle, and Sue Trostle. (Associacion De Amigos De Monteverde, 2001), 85. Eston Rockwell, "Monteverde Roads and Trucking," in *Monteverde Jubilee Family Album*, edited by Lucille Guindon, et al. (Associacion De Amigos De Monteverde, 2001), 82. **xv "the tigres had not yet been hunted . . ."** Lucille Guindon, et al., eds., "Memorable Dates," in *Monteverde Jubilee Family Album* (Associacion De Amigos De Monteverde, 2001), 260. Lucille Guindon, et al., eds., Photo Caption in *Monteverde Jubilee Family Album* (Associacion De Amigos De Monteverde, 2001), 169. **xv "To reach the type locality . . ."** Savage, "An Extraordinary New Toad (Bufo) from Costa Rica," op. cit., 153–67. **xvi "I entered into the books an orange Bufo . . ."** Norman Scott, excerpt of field notes from May 13, 1964, shared with the authors. The following descriptions of the toads in the elfin forest are also derived from these notes. **xviii "174 specimens were collected . . ."** Savage, "An Extraordinary New Toad (Bufo) from Costa Rica," op. cit., 153–67. Norman Scott, field notes, May 13, 1964. **xviii "Savage christened the toad . . ."** The species name has since been revised to *Incilius* periglenes. Savage, "An Extraordinary New Toad (Bufo) from Costa Rica," op. cit., 153–67. Twan Leenders, *Amphibians of Costa Rica* (Cornell University Press, 2016), 180. **xviii "The first pair that Savage brought back . . ."** Savage, "An Extraordinary New Toad (Bufo) from Costa Rica," op. cit., 153–67.

CHAPTER ONE: GHOST STORIES

Unless otherwise noted, details in this chapter were derived from personal communication with the following sources:

Raquel Bone, 2015
Moncho Calderon, 2015 and 2023
Eladio Cruz, 2015
Richard LaVal, 2015
Karen Masters, 2015

Paola Muñoz, 2016 and 2021
David Ritland, 2021
Adam Stein, 2016

ENDNOTES

3 "rearing tons of butterflies . . ." Alan Masters, letter to David and Robin Ritland. 5 "the last of the toads had descended . . ." Martha L. Crump, *In Search of the Golden Frog* (University of Chicago Press, 2000), 155–57. 8 "The jungle is dark . . ." Arthur Miller, *Death of a Salesman* (New York: Penguin Classics, 2000), 134. 12 "do not blame a witch . . ." David Werner, Carol Thuman, and Jane Maxwell, *Where There Is No Doctor: A Village Health Care Handbook*. Rev. ed. (Berkeley, California: Hesperian Health Guides, 2013). 12 "handbill bearing the title 'Wanted Alive.'" Conservation International, 2010. 15 "the fer-de-lance snake (Bothrops asper) . . ." Norman Scott, "*Bothrops asper* (Terciopelo, Fer-de-Lance)," in *Costa Rican Natural History*, ed. Daniel H. Janzen (University of Chicago Press, 1983), 383. 16 "When I was twenty-two . . ." Miller, op. cit., 48.

CHAPTER TWO: THE GOLDEN AGE AND THE GREEN MOUNTAIN

Unless otherwise noted, details in this chapter were derived from personal communication with the following sources:

Giovanni Bello, 2023 and 2024
Eladio Cruz, 2017, 2018, and 2021
Ricky Guindon, 2018
George Powell, 2024
Mills Tandy, 2023

Accounts of Wolf Guindon's homesteading in Monteverde are drawn in part from recorded recollections that Guindon made on cassette tapes for Kay Chornook during the writing of their book *Walking with Wolf* (2008).

21 "In the winter of 1949 . . ." Mildred and Hubert Mendenhall, 1999 letter to family and friends. Recorded in the Monteverde Jubilee Family Album, edited by Lucille Guindon, Martha Moss, Marvin Rockwell, John Trostle, and Sue Trostle (Associacion De Amigos De Monteverde, 2001), 11. 21 "nor register for the draft . . ." Marvin Rockwell, "A Brief History of Monteverde," in *Monteverde Jubilee Family Album*, edited by Lucille Guindon, et al. (Associacion De Amigos De Monteverde, 2001), 15. 21 "A supporter of the defendants replied . . ." This was Hubert Mendenhall, Clerk of the Fairhope Meeting of Friends. Mildred and Hubert Mendenhall, 1999 letter to family and friends, op. cit., 11. Marvin Rockwell, op. cit., 15. 22 "his wife was descended from . . ." Mildred and Hubert Mendenhall, 1999 letter to family and friends, op. cit., 11. 22 "plea of nolo contendere . . ." Marvin Rockwell, op. cit., 15. 22 "they served four months of their sentence . . ." Chornook and Guindon, *Walking with Wolf* (Wandering Woods Press, 2007), 21. Ibid. 22 "They ruled out frigid Canada . . ." Marvin Rockwell, op. cit., 15. 23 "particularly impressed with Costa Rica." Ibid. 23 "When the first of the expatriates . . ." Cecil F. Rockwell, "Pioneering in Costa Rica," in *Monteverde Jubilee Family Album*, edited by Lucille Guindon, et al. (Associacion De Amigos De Monteverde, 2001), 14. 23 "(the airport that Jay Savage . . ." Zúñiga, op. cit. 23 "they began a survey of land . . ." Arthur Rockwell and Eston Rockwell, *Monteverde Jubilee Family Album*, edited by Lucille Guindon, et al. (Associacion De Amigos De Monteverde, 2001), 17–19. 23 "soil ill-suited for farming." John Campbell, "Monteverde Memoirs," in *Monteverde Jubilee Family Album*, edited by Lucille Guindon, et al. (Associacion

ENDNOTES

De Amigos De Monteverde, 2001), 19–20. 23 **"news came to the group . . ."** Mildred Mendenhall, "Monteverde," in *Monteverde Jubilee Family Album,* edited by Lucille Guindon, et al. (Associacion De Amigos De Monteverde, 2001), 34. Ibid., 34–37. 24 **"Indigenous clans had seasonally occupied . . ."** Katherine Griffith, Daniel Peck, and Joseph Stuckey, "Agriculture in Monteverde: Moving toward Sustainability," in *Monteverde: Ecology and Conservation of a Tropical Cloud Forest,* edited by Nalini M. Nadkarni and Nathaniel T. Wheelwright (New York, Oxford Academic, 2000), 391. Robert M. Timm, "Prehistoric Cultures and Inhabitants," in *Monteverde: Ecology and Conservation of a Tropical Cloud Forest,* edited by Nalini M. Nadkarni and Nathaniel T. Wheelwright (New York, Oxford Academic, 2000), 408–9. 24 **"only within the last generation . . ."** Ramón Leitón and Florencia Méndez were likely the first settlers to start a farm in the San Luis Valley around 1915; Ramón had first come up on horseback from Guacimal. They planted bananas and chayote and their children found spouses at dances in Miramar, going on to establish their own homesteads in the valley. By the early 1920s, other families began to settle in San Luis and move higher up in elevation; Jose Mendes named the town of Cerro Plano when he came up from the valley in a hunting party, and by the end of the 1920s the first families had developed farms in Santa Elena and the plateau that the Quakers would soon reach on their diaspora. Accounts of these early Costa Rican settlers of the areas that would come to be known as Monteverde, Cerro Plano, and San Luis are drawn from: Carol Evans, "Life Before the Arrival of the Gringos" in *Monteverde Jubilee Family Album,* edited by Lucille Guindon, et al. (Associacion De Amigos De Monteverde, 2001), 53–4. Campbell, "Monteverde Memoirs," op. cit., 37–8. Lucille Guindon, "Early Geographic History," in *Monteverde Jubilee Family Album,* edited by Lucille Guindon, et al. (Associacion De Amigos De Monteverde, 2001), 53. 24 **"Gold had drawn the first . . ."** Leslie Burlingame, "Conservation in the Monteverde Zone: Contribution of Conservation Organizations," in *Monteverde: Ecology and Conservation of a Tropical Cloud Forest,* edited by Nalini M. Nadkarni and Nathaniel T. Wheelwright (New York, Oxford Academic, 2000), 353. 24 **"By the time the North Americans arrived . . ."** Ibid. 24 **"Arguedas, Leitón, Méndez, and Villalobos"** Mariano Arguedas, "Before the Machos Came," in *Monteverde Jubilee Family Album,* 52 (reprinted from "From Back in the Woods," vol. 2, no. 2, July 1961). Evans, op. cit., 53–4. Daisy Arguedas Méndez, "Cuando llegaron los Quakeros," in *Monteverde Jubilee Family Album,* edited by Lucille Guindon, et al. (Associacion De Amigos De Monteverde, 2001), 56. Juan Leitón Villalobos, "The Arrival of 'Los Machos'," in *Monteverde Jubilee Family Album,* edited by Lucille Guindon, et al. (Associacion De Amigos De Monteverde, 2001), 56. 24 **"selling food and guaro . . ."** Burlingame, "Conservation in the Monteverde Zone: Contribution of Conservation Organizations," op. cit., 353. 24 **"The Quakers heard stories . . ."** Roy Joe and Ruth Stuckey, "Home," in *Monteverde Jubilee Family Album,* edited by Lucille Guindon, et al. (Associacion De Amigos De Monteverde, 2001), 222. 24 **"the Quakers would pay . . ."** Campbell, "Monteverde Memoirs," op. cit., 37–8. 25 **"McDuffie had died . . ."** Mildred and Hubert Mendenhall, 1999 letter to family and friends, op. cit., 11. 25 **"Wolf joined the expatriate Quakers . . ."** Chornook and Guindon, op. cit., 27–8. 25 **"While his new wife Lucky . . ."** Ibid., 37–9. 25 **"their home was a fourteen-by-sixteen-foot . . ."** Ibid. 26 **"Wolf possessed** a claim..." Ibid., 71, 85. 26 "They were a real dull golden color..." Ibid., 70. 27 "That's just about the only time we'd see them..." Ibid. 27 "And when Jerry James set out..." Ibid., 71. 27 "The first formal research project..." Carl Rettenmeyer spent three weeks in Monteverde in February 1963 studying army ants and camp-follower insects. Roger D. Akre and Carl W. Rettenmeyer, "Behavior of Staphylinidae Associated with Army Ants (Formicidae: *Ecitonini*)." *Journal of the Kansas Entomological Society* 39, no. 4 (1966): 745–82. http://www.jstor.org/stable/25083583. Wolf Guindon, "Monteverde Beginnings," in *Monteverde: Ecology and Conservation of a Tropical Cloud Forest*, edited by Nalini M. Nadkarni and Nathaniel T. Wheelwright (New York, Oxford Academic, 2000), 10. 27 "**an ornithologist had visited...**" The ornithologist's name was Paul Slud. Lucille Guindon, et al., eds., "Memorable Dates," in *Monteverde Jubilee Family Album*, 262. 27 "**biologists compiling local species lists...**" "Monteverde Cloud Forest," Monteverde Cloud Forest, https://cloudforestmonteverde.com/. 27 "**For comparison: Yosemite National Park...**" "Nature & Science: Yosemite National Park," *National Park Service*, https://www.nps.gov/yose/learn/nature/index.htm. 27 "**The most generous estimates of the Monteverde area...**" "Monteverde Cloud Forest," https://cloudforestmonteverde.com/. "Historia," El Bosque Eterno de los Niños, https://acmcr.org/quienes-somos/historia/. 27 "**the highest diversity of orchids...**" William A. Haber, "Plants and Vegetation," in *Monteverde: Ecology and Conservation of a Tropical Cloud Forest*, edited by Nalini M. Nadkarni and Nathaniel T. Wheelwright (New York, Oxford Academic, 2000), 57. 27 "**Close to 50 percent of Costa Rica's biodiversity...**" "Monteverde Cloud Forest," https://cloudforestmonteverde.com/. 28 "**seven of Costa Rica's twelve life zones...**" Haber, op. cit., 41–4. 28 "**Savage had been instrumental in the co-founding...**" "Jay M. Savage," El Bosque Eterno de los Niños, https://bosqueternosa.wordpress.com/tag/jay-m-savage/. Jay M. Savage, "Preface," op. cit., xi. Leslie J. Burlingame, "Evolution of the Organization for Tropical Studies," *Revista de Biología Tropical* 50 no. 2 (2002): 439–72. Oscar J. Rocha and Elizabeth Braker, "The Organization for Tropical Studies: History, Accomplishments, Future Directions in Education and Research, with an Emphasis in the Contributions to the Study of Plant Reproductive Ecology and Genetics in Tropical Ecosystems," *Biological Conservation* 253 (2021): https://doi.org/10.1016/j.biocon.2020.108890. 28 "**the first group slept in the loft...**" John Campbell, "Biologists," in *Monteverde Jubilee Family Album*, edited by Lucille Guindon, et al. (Associacion De Amigos De Monteverde, 2001), 169. 28 "**George Powell's report on local bird species...**" Burlingame, "Conservation in the Monteverde Zone: Contribution of Conservation Organizations," op. cit., 355. 29 "**Ornithologists drove their pickup trucks...**" The ornithologist in question was Peter Feinsinger, who made the trip in September 1971. "The Early Biologists and Naturalists of Monteverde," *Bosqueterno S.A.*, February 1, 2010, https://bosqueternosa.wordpress.com/2010/02/01/the-early-biologists-and-naturalists-of-monteverde/. 29 "**entomologists arrived to study mimicry...**" This was Bill Haber's experience. "The Early Biologists and Naturalists of Monteverde," *Bosqueterno S.A.* 29 "**mammalogists adopted nocturnal...**" Richard LaVal would become one of Monteverde's most accomplished bat researchers; he arrived from the University of Kansas in 1973. "The Early Biologists and Naturalists of Monteverde," *Bosqueterno S.A.* 29 "**canopy**

ENDNOTES

researchers climbed into the treetops..." Canopy researcher Nalini Nadkarni reached the mountain town in 1979. Peter Beaumont, "Head in the Clouds: The Woman Scaling Fresh Climate Change Heights," *The Guardian* 2018, https://www.theguardian.com/global-development/2018/dec/16/head-in-the-clouds-climate-change-nalini-nadkarni-costa-rica-monteverde. 29 "Under the shade of a tall..." Campbell, "Biologists," op. cit., 169. 29 "bamboo pole and mist net..." Chornook and Guindon, op. cit., 84. 29 "Then the terrible sound started up again..." George Powell, "The Monteverde Preserve's Beginning: A Couple of Frequently Asked Questions," *Monteverde Jubilee Family Album* (Associacion De Amigos De Monteverde, 2001), 170. 30 "He was not far at all..." Chornook and Guindon, op. cit., 84. 30 "Al suelo!" Ibid., 51. 30 "Ironically, the first time George Powell..." Ibid., 83. 30 "when a professor at the University of Costa Rica..." Bruce E. Young and David B. McDonald, "Birds," *Monteverde: Ecology and Conservation of a Tropical Cloud Forest*, edited by Nalini M. Nadkarni and Nathaniel T. Wheelwright (New York, Oxford Academic, 2000), 179. 30 "They arrived with their wives..." Campbell, "Biologists," op. cit., 169. 30 "they selected one of the only large forest plots..." Ibid. 31 "John Campbell took the very first photograph..." Lucille Guindon, et al., eds., "Memorable Dates," in *Monteverde Jubilee Family Album.* 31 "Powell had approached Wolf Guindon..." Chornook and Guindon, op. cit., 84. 32 "the first Homelite chain saw dealer..." Ibid., 62. Powell, op. cit., 170. 32 "a farm in the middle of the golden toad's..." Chornook and Guindon, op. cit., 85–6. 32 "the ridgeline forest was not the perfect site..." Ibid., 86. 32 "He and his sons had spent weeks..." Ibid. 32 "His new neighbor George Powell..." Ibid., 85–7. 32 "There was a need for working farms..." Ibid., 85. 33 "Wolf was open to the offer." Ibid., 86. 33 "working on behalf of land speculators..." Ibid. 33 "three major access trails." Powell, op. cit., 170. Burlingame, "Conservation in the Monteverde Zone: Contribution of Conservation Organizations," op. cit., 357. 33 "two homesteaders had begun to widen the trail..." These were the brothers Edwin and Bolivar Fonseca. Chornook and Guindon, op. cit., 81, 90. 33 "Wolf agreed not to sell..." Ibid., 86. 33 "Powell asked Wolf for help..." The second landowner was Luis Chavarría. Ibid. 33 "Powell had to call in a favor..." The professors were Douglas Robinson and Sergio Salas. Powell, op. cit., 170. 34 "five properties in the ridgeline elfin forest..." Chornook and Guindon, op. cit., 86. 34 "I got involved with George..." Ibid., 88. 35 "Wolf had brought in a trusted partner..." "The Early Biologists and Naturalists of Monteverde," *Bosqueterno S.A.* Ibid., 88. 35 "they spent long nights in a refugio..." Ibid., 88–9. 35 "the young Tico had heard..." Eladio Cruz, "Libro Vivo" speaker series, hosted by the Monteverde Institute and recorded by the authors in 2017. 35 "when Eladio was twelve years old..." Ibid. 36 "Eladio stumbled into a cacique-fueled fiesta..." Ibid. 36 "Eladio would go on to build..." Chornook and Guindon, op. cit., 108. 37 "George and Harriett Powell had invested their savings..." Burlingame, "Conservation in the Monteverde Zone: Contribution of Conservation Organizations," op. cit., 357. Chornook and Guindon, op. cit., 88. 37 "Harriett typed letters..." Chornook and Guindon, op. cit., 88. 37 "George sent the word out..." Burlingame, "Conservation in the Monteverde Zone: Contribution of Conservation Organizations," op. cit., 358. Chornook and Guindon, op. cit., 88. 37 "the

ENDNOTES

World Wildlife Fund was interested . . ." Burlingame, "Conservation in the Monteverde Zone: Contribution of Conservation Organizations," op. cit., 357. 38 "The local Costa Rican subsistence farmers . . ." Timm, op. cit., 408–9. 38 "the community set aside almost a third . . ." Burlingame, "Conservation in the Monteverde Zone: Contribution of Conservation Organizations," op. cit., 356. 38 "when John and Doris Campbell established their farm . . ." Campbell, "Biologists," op. cit., 169. "The Early Biologists and Naturalists of Monteverde," *Bosqueterno S.A.* 38 "Monteverde's unofficial weather man . . ." "The Early Biologists and Naturalists of Monteverde," *Bosqueterno S.A.* John Campbell, "Weather," in *Monteverde Jubilee Family Album,* edited by Lucille Guindon, et al. (Associacion De Amigos De Monteverde, 2001), 172. 38 "Mary taught biology classes . . ." "The Early Biologists and Naturalists of Monteverde," *Bosqueterno S.A.* 38 "George Powell transferred the 328 hectares . . ." Chornook and Guindon, op. cit., 88–9. 38 "Guindon signed on as a part-time employee . . ." Ibid., 90. 39 "George and Harriett returned to the United States . . ." Lucille Guindon, et al., eds., "Memorable Dates," in *Monteverde Jubilee Family Album*, 265. Ibid., 91. 39 "Mills Tandy had traveled to Monteverde . . ." Mills Tandy, "Reproductive Behavior and Possible Causes for Decline of the Apparently Extinct Golden Toad, *Bufo periglenes*," *Images of the Natural World,* https://imagesofthenaturalworld.wordpress.com. 41 "those images had been taken at an artificial site . . ." Ibid. 41 "With help from Wolf Guindon and Giovanni Bello . . ." Chornook and Guindon, op. cit., 99. 41 "Tandy had selected eight study sites . . ." Tandy, op. cit. 41 "he counted 164 toads . . ." Ibid. 41 "The roots had grown around . . ." Ibid. 42 "When Michael and Patricia Fogden arrived . . ." "Michael and Patricia Fogden," Minden Pictures: https://www.mindenpictures.com/favorites/photographer-portfolios/michael-and-patricia-fogden.html. 42 "To reach the area where the toads live . . ." Details and quotations are drawn from Michael and Patricia Fogden's 1984 article in *Natural History*: Michael Fogden and Patricia Fogden, "All That Glitters May Be Toads," *Natural History*: 93 (1984), 46–50. 43 "By 1981, the old oxcart road . . ." Kent Britt, "Costa Rica Steers the Middle Course," *National Geographic,* July 1981, 47. 43 "Savior of a primeval forest. . ." Ibid. 43 "with thirteen new national parks . . ." Ibid. 43 "The Metro Toronto Zoo . . ." Chornook and Guindon, op. cit., 76. 43 "the Reserve kept a few toads in a terrarium . . ." Ibid., 75. 44 "Irma Rockwell once called the Reserve . . ." Ibid., 76. 44 "Mills Tandy was at a conference . . ." Tandy, op. cit. 44 "Wolf Guindon heard about advertisements . . ." Chornook and Guindon, op. cit., 79. 45 "The foundations of Monteverde . . ." Burlingame, "Conservation in the Monteverde Zone: Contribution of Conservation Organizations," op. cit., 373. 46 "Wolf Guindon fought the wind . . ." Chornook and Guindon, op. cit., 75.

ENDNOTES

CHAPTER THREE: THE END OF THE SHOW

Unless otherwise noted, details in this chapter were derived from personal communication with the following sources:

Martha Crump, 2018	Frank Hensley, 2023
Hazel Guindon, 2018	Priscilla Palavicini, 2018 and 2024
Ricky Guindon, 2018	Nicole Rockwell, 2018

56 **"not yet sold off as adventure parks . . ."** As tourism blossomed in Monteverde, some families sold their farms to purchase taxis or to the developers of adventure parks offering hanging bridges, bungee jumping, and zip-lining. 58 **"Martha Crump watched the storm break . . ."** Details from Martha Crump's first encounter with the golden toads are drawn from her field notes published in her book *In Search of the Golden Frog* (University of Chicago Press, 2000) and a personal conversation with Trevor Ritland and Priscilla Palavicini in October 2018. 58 **"It had been raining for two days . . ."** Crump, op. cit., 149. 58 **"Crump had been in Monteverde since March . . ."** Ibid. 58 **"the golden toads are out!"** Ibid. 59 **"Marty, you're not going to believe it!"** Ibid. 59 **"Marty Crump met Wolf . . ."** Ibid., 150. 59 **"jewels scattered about the dim understory . . ."** Ibid. 60 **"She told Wolf that she planned . . ."** Ibid. 60 **"Wolf had volunteered to assist . . ."** Details of Vandenberg's early attempts to study *Bufo periglenes* are drawn from Wolf Guindon's recorded memories in Chornook and Guindon's *Walking with Wolf*, 73–4. 61 **"It wasn't elfin woods or an exposed ridge . . ."** Ibid., 74. 61 **"Vandenberg's 1972 study didn't pan out . . ."** Ibid., 73–4. 61 **"Vandenberg returned to undertake . . ."** The following details from Vandenberg's study are drawn from John J. Vandenberg and Susan K. Jacobson, "Reproductive Ecology of the Endangered Golden Toad (*Bufo periglenes*)," *Journal of Herpetology* 25, no. 3 (Sept. 1991): 321–7. 62 **"a release call performed . . ."** Ibid. 63 **"No frogs marked in 1977 were observed . . ."** Ibid. Susan K. Jacobson, "Short Season of the Golden Toad," *International Wildlife*, November/December 1983, 26. 63 **"Susan Jacobson published her observations . . ."** Jacobson, op. cit., 26. 63 **"a lifespan of around ten years . . ."** Andrew R. Blaustein, "Chicken Little or Nero's Fiddle? A Perspective on Declining Amphibian Populations," *Herpetologica* 50, no. 1 (1994), 85–97. 63 **"Wolf returned to the El Valle site . . ."** Walter Timmerman, at that time the director of the Reserve, accompanied Wolf on this excursion. Chornook and Guindon, op. cit., 74–5. 64 **"Beginning April 8, Marty Crump . . ."** Crump, op. cit., 150–3. 64 **"dried and desiccated eggs . . ."** Ibid., 153. 64 **"the golden toads reemerged briefly . . ."** Ibid., 153–6. 65 **"If the mist and drizzle continue . . ."** Ibid., 155. 65 **"her final field entries for the 1987 season noted . . ."** Ibid., 157. 65 **"In 1988, the spring was drier . . ."** Ibid. 65 **"In early May, Marty Crump returned . . ."** Ibid., 158. 65 **"He counted nine toads . . ."** Ibid., 160. 65 **"she discovered one male toad . . ."** Ibid., 158. 66 **"Even as the cold and rain persisted . . ."** Ibid., 158–61. 66 **"I'm thinking that the window of opportunity . . ."** Ibid., 161. 66 **"I hop yor goldin tods come out . . ."** Ibid. 66 **"the alarm clock of a troop of howler monkeys . . ."** Ibid. 66 **"one of Crump's graduate students . . ."** Ibid., 163. 67 **"Crump received a distressing call . . ."** Hensley notes that he collaborated with the biologist Cathy Langtimm to help cover the responsibilities

… ENDNOTES

of Crump's fieldwork in her absence. Ibid., 164. 74 **"There were no reports at all of dead or dying golden toads..."** Ibid., 158–61.

CHAPTER FOUR: THE CREEPING FEAR

Unless otherwise noted, details in this chapter were derived from personal communication with the following sources:

    Lee Berger, 2023 and 2024   Joyce Longcore, 2023 and 2024
    Martha Crump, 2018   Allan Pessier, 2023 and 2024
    Greg Czechura, 2023

79 **"the First World Congress of Herpetology"** Michael R. K. Lambert, "First World Congress of Herpetology, Held at the University of Kent, Canterbury, England, UK, during 11–19 September 1989." Environmental Conservation 17, no. 1 (1990): 85–6. https://doi.org/10.1017/S037689290001746X. 79 **"Marty Crump was not scheduled..."** Kathryn Phillips, *Tracking the Vanishing Frogs: An Ecological Mystery* (Penguin Mass Market, 1994), 21. 80 **"she told a few friends..."** Crump, op. cit., 172. 80 **"Cynthia Carey, a physiological ecologist..."** James P. Collins and Marty Crump, *Extinction in Our Times: Global Amphibian Decline* (Oxford University Press, 2009): 137–8. 80 **"Stanley Rand, from the Smithsonian..."** W. Ronald Heyer, A. Stanley Rand, Carlos Alberto Goncalvez da Cruz, and Oswaldo L. Peixoto, "Decimations, Extinctions, and Colonizations of Frog Populations in Southeast Brazil and Their Evolutionary Implications," *Biotropica* 20, no. 3 (Sept. 1988): 230–5. W. Ronald Heyer, "Obituary: In Celebration of A. Stanley Rand," Smithsonian Institute. *Phyllomedusa* 5, no. 1 (2006): 8–10. 80 **"Tyler had been among the biologists..."** Greg Roberts, "Reflecting on Campaign to Save the Conondale Range," *Sunshine Coast Birds*, September 28, 2016. https://sunshinecoastbirds.blogspot.com/2016/09/reflecting-on-campaign-to-save.html. 80 **"swallowed her own fertilized eggs..."** Chris J. Corben, Glen J. Ingram, and Michael J. Tyler, "Gastric Brooding: Unique Form of Parental Care in an Australian Frog," *Science* 186 no. 4167 (Dec. 1974): 946–7. https://doi.org/10.1126/science.186.4167.946. 80 **"gastric brooding was so unbelievable..."** Roberts, op. cit. 80 **"one of Tyler's contemporaries was forced..."** Greg V. Czechura and Glen J. Ingram, "*Taudactylus diurnus* and the Case of the Disappearing Frogs," *Memoirs of the Queensland Museum* 29, no. 2 (1990): 361–5. 81 **"last sighting of the day frog..."** Glen J. Ingram and Keith R. McDonald, "An Update on the Decline of Queensland's Frogs." *Herpetology in Australia: A Diverse Discipline.* D. Lunney and D. Ayers, eds., *Transactions of the Royal Zoological Society of New South Wales* (1993): 297–303. 81 **"the pine forests near Oaxaca..."** Marcia Barinaga, "Where Have All the Froggies Gone?" *Science* 247, no. 4946 (1990): 1033–4. DOI:10.1126/science.247.4946.1033. 81 **"disappearance of the mountain yellow-legged frogs..."** Cathy Brown, Marc Hayes, Gregory Green, and Diane Macfarlane, "Mountain Yellow-Legged Frog Conservation Assessment for the Sierra Nevada Mountains of California, USA," USDA Forest Service, California Department of Fish and Wildlife, National Park Service, U.S. Fish and Wildlife Service (July 2014): 30. https://amphibiaweb.org/refs/pdfs/USFS_RAMU_CA_508_final.pdf. 81 **"Marc Hayes and Mark Jennings..."**

ENDNOTES

Marc P. Hayes and Mark R. Jennings, "Decline of Ranid Frog Species in Western North America: Are Bullfrogs (*Rana catesbeiana*) Responsible?" *Journal of Herpetology* 20, no. 4 (1986): 490–509. 81 **"Wake heard from David Bradford . . ."** Andrew Blaustein and David Wake, "Declining Amphibian Populations: A Global Phenomenon?" *Tree* 5, no. 7 (1990): 203–4. Phillips, op. cit., 114–6. 81 **"Wake had returned to the Sierra Nevada. . ."** Timothy R. Halliday, "The Case of the Vanishing Frogs," *Technology Review*, 1997. https://www.technologyreview.com/1997/05/01/237275/the-case-of-the-vanishing-frogs/. 81 **"When Wake heard rumors . . ."** J. Alan Pounds, "Monteverde Salamanders, Golden toads, and the Emergence of the Global Amphibian Crisis," in Monteverde: Ecology and Conservation of a Tropical Cloud Forest, 172. Collins and Crump, op. cit., 27. 82 **"Wake had visited Monteverde to study . . ."** Pounds, "Monteverde Salamanders, Golden Toads, and the Emergence of the Global Amphibian Crisis," op. cit., 172. 82 **"Crump and her graduate student Alan Pounds had counted . . ."** Martha L. Crump and J. Alan Pounds, "Lethal Parasitism of an Aposematic Anuran (*Atelopus varius*) by *Notochaeta bufonivora* (Diptera: Sarcophagidae)," *The Journal of Parasitology* 71, no. 5 (Oct. 1985): 588–91. Crump, op. cit., 160. 82 **"they had found as many as two hundred . . ."** Crump and Pounds, "Lethal Parasitism of an Aposematic Anuran (*Atelopus varius*) by *Notochaeta bufonivora* (Diptera: Sarcophagidae)," op. cit., 588. 82 **"she searched for six hours . . ."** Crump, op. cit., 160. 82 **"'What's happened?' she wrote . . ."** Ibid. 83 **advertised in European pet trade magazines . . .** Ibid., 160–1. 83 **"But when could they have done this?"** Ibid., 161. 83 **"a particularly nasty parasite . . ."** Ibid., 160. 83 **"He opened the box to find . . ."** Donald K. Nichols, "Tracking Down the Killer Chytrid of Amphibians." *Herpetological Review* 34, no. 2 (June 2003): 101–4. 84 **"numerous microscopic single-celled organisms unlike anything I had ever seen before."** Ibid., 102. 84 **"Because I am looking for something . . ."** Ibid., 103. 84 **"Where Have All the Froggies Gone?"** Barinaga, op. cit., 1033–4. 84 **"The Case of the Disappearing Frogs."** Czechura and Ingram, op. cit., 361–5. 84 **"Why Are Frogs Croaking?"** William F. Laurance, "Why are Queensland's Frogs Croaking?" *Nature Australia* 25 (1996): 56–62. 85 **"acid rain, pesticide poisoning . . ."** Blaustein and Wake, op. cit., 204. 85 **"harmful UV radiation from ozone depletion . . ."** Barinaga, op. cit., 1034. 85 **"he'd been working on a presentation . . ."** Donald K. Nichols, Anthony J. Smith, and Chris H. Gardiner, "Dermatitis of Anurans Caused by Fungal-like Protists," *Proceedings of the American Association of Zoo Veterinarians* (1996): 220. 87 **"one of the world's leading experts on chytrids . . ."** D. Rabern Simmons, "Biography of Dr. Joyce E. Longcore," *Collection of Zoosporic Eufungi at University of Michigan* (CZEUM). https://czeum.herb.lsa.umich.edu/biography-of-dr-joyce-e-longcore-by-d-rabern-simmons/. 88 **"New Culprit in Deaths of Frogs."** "New Culprit in the Deaths of Frogs," *The New York Times*, September 16, 1997. https://www.nytimes.com/1997/09/16/science/new-culprit-in-deaths-of-frogs.html. 89 **"The sliver of a crescent moon . . ."** "Complete Sun and Moon Data for One Day: Fortuna, N 08.7263°, W 82.242°, Friday, 1997-January-10," *Astronomical Applications Department of the U.S. Naval Observatory*, https://aa.usno.navy.mil/calculated/rstt/oneday?date=1997-01-10&lat=08.7263&lon=-82.242&label=Fortuna&tz=6&tz_sign=-1&tz_label=false&dst=false&submit=Get+Data. 89 **"the long-poled net she carried . . ."**

ENDNOTES

Karen Lips, "What If There Is No Happy Ending? Science Communication as a Path to Change," *Scientific American* (May 2013). 89 "But on this January morning in 1997, Lips . . ." Ibid. 90 "Her adviser at the University of Miami . . ." Perspectives in Conservation. *Herpetological Review* 50, no. 2 (2019): 311–4. 90 "the little frogs looked like green and gray lichen . . ." Karen R. Lips, "Witnessing Extinction in Real Time," *PLoS Biology* 16, no. 2 (2018). https://doi.org/10.1371/journal.pbio.2003080. 90 "I was living a field biologist's dream . . ." Ibid. 90 "In December 1992 . . ." Lips, "What If There Is No Happy Ending? Science Communication as a Path to Change," op. cit. 90 "she soothed her concerns with rationalizations . . ." Lips, "Witnessing Extinction in Real Time," op. cit. 90 "they would jump once, and then expire . . ." Brendan Bane, "Karen Lips, herpetologist," *UC Santa Cruz - Science Communication Master's Program* (April 2016). https://scicom.ucsc.edu/publications/QandA/2016/lips.html. 90 "She made a list of theories . . ." Collins and Crump, op. cit., 146. 91 "She had heard about the golden toads . . ." Lips, "Witnessing Extinction in Real Time," op. cit. 91 "she moved southeast . . ." Ibid. 91 "frozen in their normal calling postures . . ." Karen R. Lips, "Mass Mortality and Population Declines of Anurans at an Upland Site in Western Panama," *Conservation Biology* 13, no. 1 (1999): 117. https://doi.org/10.1046/j.1523-1739.1999.97185.x. 92 "he merely sat there in her open palm . . ." Lips, "Witnessing Extinction in Real Time," op. cit. 93 "I could find no evidence . . ." "New Culprit in the Deaths of Frogs," *The New York Times,* op. cit. 93 "This article even has a photomicrograph . . ." Don Nichols, personal email to Joyce Longcore, September 25, 1997. 94 "Berger had been working as a PhD student . . ." Colin Ward, "Lee Berger," *CSIROpedia*. September 16, 2014. https://csiropedia.csiro.au/berger-lee/. 94 "the last to see the southern day frog . . ." Czechura and Ingram, op. cit., 363. 95 "The Mystery of the Disappearing Frog," Glen J. Ingram, "The Mystery of the Disappearing Frogs." Wildlife Australia 27, no. 3 (1990): 6–7. 95 "The Twilight Zone." Greg Czechura, "The Twilight Zone." *Wildlife Australia* 28 (1991): 19–22. 95 "For three years, McDonald had been combing . . ." Ingram and McDonald, op. cit., 297–303. Stephen J. Richards, Keith R. McDonald, and Ross A. Alford, "Declines in Populations of Australia's Endemic Tropical Rainforest Frogs," *Pacific Conservation Biology* 1, no. 1 (1994): 66–77. https://doi.org/10.1071/PC930066. 95 "McDonald partnered up with Rick Speare." "Amphibian Disease Detectives: The History of Chytridiomycosis (*Bd*) Discovery in Australia (1993–1999)." *James Cook University, Taronga Conservation Society, Australian Registry of Wildlife Health*. https://arwh.org/wp-content/uploads/2021/03/The_History_of_Chytrid_discovery_in_Australia_1993-1999.pdf. 95 "A veterinarian and medical doctor. . ." Lee Berger, Lee F. Skerratt, Ian Beveridge, and David M. Spratt, "In Memory of Rick Speare." *EcoHealth* 13 (2016): 435–7. https://doi.org/10.1007/s10393-016-1151-7. 95 "the two men began to suspect an infectious disease . . ." Lee Berger, Alexandra A. Roberts, Jamie Voyles, Joyce E. Longcore, Kris A. Murray, and Lee F. Skerratt, "History and Recent Progress on Chytridiomycosis in Amphibians." *Fungal Ecology* 19 (2016): 89–99. https://doi.org/10.1016/j.funeco.2015.09.007. 95 "As early as 1989, Rick Speare . . ." Judith Maunders, "Silent Streams." *Ecos* 109 (October–December 2001): 8–10. 95 "In 1994, Speare collaborated in the

ENDNOTES

founding . . ." "Amphibian Disease Detectives: The History of Chytridiomycosis (*Bd*) Discovery in Australia (1993–1999)." James Cook University, Taronga Conservation Society, Australian Registry of Wildlife Health. **95 "a paper published with McDonald . . ."** William F. Laurance, Keith R. McDonald, and Richard Speare, "In Defense of the Epidemic Disease Hypothesis." *Conservation Biology* 11, no. 4 (August 1997): 1030–4. **97 "It was George Rabb . . ."** Perspectives in Conservation. *Herpetological Review* 50, no. 2 (2019): 311–4. **98 "Lee Berger published her dissertation . . ."** Lee Berger, et al., "Chytridiomycosis Causes Amphibian Mortality Associated with Population Declines in the Rain Forests of Australia and Central America." *Proceedings of the National Academy of Sciences*, 95 no. 15 (1998): 9031–6. https://doi.org/10.1073/pnas.95.15.9031. **98 "Joyce Longcore, Don Nichols, and Allan Pessier published their findings . . ."** Joyce E. Longcore, Allan P. Pessier, and Donald K. Nichols, "*Batrachochytrium* dendrobatidis gen. et sp. nov., a Chytrid Pathogenic to Amphibians." *Mycologia* 91, no. 2 (1999): 219–27. https://www.jstor.org/stable/3761366.

CHAPTER FIVE: IN THE FORESTS OF THE LOST FROGS

Unless otherwise noted, details in this chapter were derived from personal communication with the following sources:

Forrest Brem, 2024
Federico Bolaños, 2018
Joyce Longcore, 2024

Karen Masters, 2024
Alan Pounds, 2018
Adam Stein, 2024

103 "Three hundred Fleischmann's glass frogs . . ." Phillips, op. cit., 141. Marc P. Hayes, "Predation on the Adults and Prehatching Stages of Glass Frogs *(Centrolenidae)*." *Biotropica* 15, no. 1 (March 1983): 74–6. https://www.jstor.org/stable/2388005. **103 "news had reached Pounds . . ."** J. Alan Pounds, "Disappearing Gold." *BBC Wildlife* 8, no. 12 (December 1990): 814. **103 "So working closely with the local photographer . . ."** The survey area was an east-to-west belt of forest straddling the Continental Divide, fifteen kilometers long and two kilometers wide. The area included habitats in each of Monteverde's climate and vegetation zones, and had historically contained populations of all of the species they were looking for. Using data from earlier studies, Pounds and his collaborators determined that normal population dynamics and environmental changes could not account for the disappearances in Monteverde: something larger was at work in the amphibian declines. Unless otherwise noted, details and methods of the 1990–1994 Monteverde amphibian surveys are drawn from the Pounds et al. 1997 *Conservation Biology* paper (citation below): J. Alan Pounds, Michael P. L. Fogden, Jay M. Savage, and George C. Gorman, "Tests of Null Models for Amphibian Declines on a Tropical Mountain." *Conservation Biology* 11, no. 6 (1997): 1307–22. http://www.jstor.org/stable/2387358. **103 "Pounds published a story in BBC Wildlife . . ."** Unless otherwise noted, details in this section are drawn from Pounds's 1990 *BBC Wildlife* article: Pounds, "Disappearing Gold." op. cit., 813–7. **104 "Alan Pounds and Marty Crump reunited . . ."** Unless otherwise noted, details in this section are drawn from the 1994 Pounds and Crump *Conservation Biology* paper: J. Alan Pounds and Martha L. Crump, "Amphibian Declines and Climate Disturbance: The Case of the Golden Toad and the Harlequin Frog." *Conservation Biology* 8 (1994): 72–85. **104 "The data lined up with Marty Crump's own observations . . ."** Crump, op. cit., 157. **104 "Pounds and Crump proposed the 'climate-linked epidemic hypothesis' . . ."** Pounds and Crump actually proposed two theories in this paper; the other expounded on earlier hypotheses that airborne contaminants like pesticides had reached critical concentrations during the warm-and-dry conditions. Pounds and Crump, "Amphibian Declines and Climate disturbance: The Case of the Golden Toad and the Harlequin Frog," op. cit., 72–85. Collins and Crump, op. cit., 145. **104 "Pounds would spend the next five years . . ."** In 1999, Pounds collaborated with Fogden and Campbell to develop these hypotheses in their paper: J. Alan Pounds, Michael P.L. Fogden, and John H. Campbell, "Biological Response to Climate Change on a Tropical Mountain." *Nature* 398 (1999): 611–5. https://doi.org/10.1038/19297. **104 "a constellation of demographic changes. . ."** Ibid. **105 "The papers from the collaborators at Urbana . . ."** Berger et al., "Chytridiomycosis causes amphibian mortality associated with population declines in the rain forests of Australia and Central America," op. cit., 9031–6. Longcore, Pessier, and Nichols, op. cit., 219–27. **105 "the theoretical pathogen . . ."** Pounds and Crump, "Amphibian Declines and Climate Disturbance: The Case of the Golden Toad and the Harlequin Frog," op. cit., 72–85. **105 "Pounds acknowledged Batrachochytrium dendrobatidis . . ."** J. Alan Pounds, "Amphibians and Reptiles," in *Monteverde: Ecology and Conservation of a Tropical Cloud Forest,* edited by Nalini M. Nadkarni and Nathaniel T. Wheelwright (New York, Oxford Academic, 2000): 161. J. Alan Pounds and Robert Puschendorf, "Clouded Futures." *Nature* 427 (2004): 107–9. **105 "At least one other factor . . ."** Pounds, "Amphibians and Reptiles," op. cit., 162. **105 "Extreme changes in moisture and temperature . . ."** Pounds, "Amphibians and Reptiles," op. cit., 162. Pounds and Crump, "Amphibian Declines and Climate Disturbance: The Case of the Golden Toad and the Harlequin Frog," op. cit., 72–85. **105 "If the chytrid fungus was the bullet . . ."** Robin Moore, *In Search of Lost Frogs* (Firefly Books, 2014), 62. Mattha Busby, "'The Ghost That Haunts Monteverde': How the Climate Crisis Killed the Golden Toad." *The Guardian*, November 21, 2022. https://www.theguardian.com/environment/2022/nov/21/golden-toad-haunts-monteverde-how-species-foretold-climate-crisis-aoe. **106 "Fungi were present in our seas . . ."** "The Great Oxidation Event: How Cyanobacteria Changed Life." American Society for Microbiology, February 18, 2022. https://asm.org/articles/2022/february/the-great-oxidation-event-how-cyanobacteria-change. **106 "By some accounts, it was a fungus that first crept . . ."** Daniel S. Heckman, David M. Geiser, Brooke R. Eidell, Rebecca L. Stauffer, Natalie L. Kardos, and S. Blair Hedges, "Molecular Evidence for the Early Colonization of Land by Fungi and Plants." *Science* 293 (2001): 1129–33. DOI:10.1126/science.1061457. **106 "Around the time of the Cambrian explosion . . ."** T. N. Taylor, S. D. Klavins, M. Krings, E. L. Taylor, H. Kerp, and H. Hass, "Fungi from the Rhynie chert: A view from the Dark Side," *Transactions of the Royal Society of Edinburgh: Earth Sciences*, 94, no. 4 (2003): 457–73. https://doi.org/10.1017/S026359330000081x. T. Taylor, W. Remy, and H. Hass, "Allomyces in the Devonian." *Nature* 367 (1994): 601. https://doi.org/10.1038/367601a0. **106 "When the volcanic eruptions . . ."** Mark A. Sephton, Henk

ENDNOTES

Visscher, Cindy V. Looy, Alexander B. Verchovsky, and Jonathan S. Watson, "Chemical Constitution of a Permian-Triassic Disaster Species," *Geology* 37, no. 10 (2009): 875–8. doi: https://doi.org/10.1130/G30096A.1. 106 **"when fungi dominated the planet."** Yoram Eshet, Michael R. Rampino, and Henk Visscher, "Fungal Event and Palynological Record of Ecological Crisis and Recovery across the Permian-Triassic Boundary." *Geology* 23, no. 11 (1995): 967–70. doi: https://doi.org/10.1130/0091-7613(1995)023<0967:FEAPRO>2.3 .CO;2. 106 **"the chytrids were there . . ."** Arturo Casadevall, "Fungi and the Rise of Mammals." *PLoS Pathog* 8, no. 8 (2012). https://doi.org/10.1371/journal. ppat.1002808, 106 **"From the crater of the K-T impact . . ."** Yan-Jie Feng, et al., "Phylogenomics Reveals Rapid, Simultaneous Diversification of Three Major Clades of Gondwanan Frogs at the Cretaceous–Paleogene Boundary." *Proceedings of the National Academy of Sciences*, 114, no. 29 (2017). https://doi.org/10.1073/pnas.1704632114. 107 **"in the jungles and ponds . . ."** Simon J. O'Hanlon, et al., "Recent Asian Origin of Chytrid Fungi Causing Global Amphibian Declines." *Science* 360 (2018): 621–7. DOI:10.1126/science .aar1965. 107 **"as early as 1915 in Southern California . . ."** A. J. Adams, A. P. Pessier, and C. J. Briggs, "Rapid Extirpation of a North American Frog Coincides with an Increase in Fungal Pathogen Prevalence: Historical Analysis and Implications for Reintroduction." *Ecology and Evolution* 7, no. 23 (2017): 10216–32. https://doi.org/10.1002/ece3.3468. 107 **"and 1926 in Baja . . ."** Andrea J. Adams, Anny Peralta-García, Carlos A. Flores-López, Jorge H. Valdez-Villavicencio, and Cheryl J. Briggs, "High Fungal Pathogen Loads and Prevalence in Baja California Amphibian Communities: The Importance of Species, Elevation, and Historical Context." *Global Ecology and Conservation* 33 (2022). https://doi.org/10.1016/j .gecco.2021.e01968. 107 **"*Bd*ASIA-1 in Korea, *Bd*BRAZIL in South America, and *Bd*CAPE in South Africa."** O'Hanlon et al., "Recent Asian Origin of Chytrid Fungi Causing Global Amphibian Declines," op. cit., 621–7. 107 **"two strains of genetically isolated populations . . ."** Rhys A. Farrer, et al. (2011). Multiple Emergences of Genetically Diverse Amphibian-Infecting Chytrids Include a Globalized Hypervirulent Recombinant Lineage. Proceedings of the National Academy of Sciences 108, no. 46, 18732–6. https://doi .org/10.1073/pnas.1111915108. 108 **"scientist named Lancelot Hogben."** George Philip Wells, "Lancelot Thomas Hogben, 9 December 1895–22 August 1975." *Biogr. Mems Fell. R. Soc* 24 (1997): 183–221. http://doi.org/10.1098/rsbm.1978.0007. Ed Yong, "How a frog became the first mainstream pregnancy test," *The Atlantic* (May 2017). https://www.theatlantic.com /science/archive/2017/05/how-a-frog-became-the-first-mainstream-pregnancy-test /525285/. 108 **"he and his wife were once reported . . ."** Wells, op. cit., 197. 109 **"supplies seem to be unlimited and export unrestricted."** Edward R. Elkan, "The xenopus pregnancy test." *The British Medical Journal* 2, no. 1 (1938): 1253–74. https://doi.org/10.1136 /bmj.2.4067.1253. 109 **"a hundred frogs a month."** Alex Casey, "*We need to talk about how frogs used to be pregnancy tests.*" *The Spinoff*, June 2022. https://thespinoff.co.nz /science/17-06-2022/we-need-to-talk-about-how-frogs-used-to-be-pregnancy-tests. 109 **"quite tricky to open without allowing a frog stampede . . ."** Lance van Sittert and G. John Measey, "*Historical perspectives on global exports and research of African clawed frogs (Xenopus laevis).*" *Transactions of the Royal Society of South Africa* 71, no. 2 (2016): 157–66.

ENDNOTES

https://doi.org/10.1080/0035919X.2016.1158747. 109 "the world's most widely distributed amphibian." Sittert and Measey, op. cit., 157–66. 109 "Well adapted to and largely unbothered . . ." Rachel Nuwer, "Doctors Used to Use Live African Frogs as Pregnancy Tests." *Smithsonian.com*, May 16, 2013. https://www.smithsonianmag.com/smart-news/doctors-used-to-use-live-african-frogs-as-pregnancy-tests-64279275/. 109 "By 1978, it had reached a port in Brisbane . . ." Lee Berger and Lee F. Skerratt, "Disease Strategy *Chytridiomycosis* (Infection with *Batrachochytrium dendrobatidis*)," Department of Sustainability, Environment, Water, Populations and Communities, Public Affairs, Canberra, ACT, Australia, 2012. http://www.environment.gov.au/biodiversity/invasive-species/publications/preparation-disease-strategy-manual-amphibian-chytrid-fungus/. 109 "in the early 1970s, it had arrived in Mexico." Tina L. Cheng, Sean M. Rovito, David B. Wake, and Vance T. Vredenburg, "Coincident Mass Extirpation of Neotropical Amphibians with the Emergence of the Infectious Fungal Pathogen *Batrachochytrium dendrobatidis*," *Proc Natl Acad Sci USA* 108, no. 23 (June 2011): 9502-7. doi: 10.1073/pnas.1105538108. 109 "Sometime in the 1980s, it struck Guatemala . . ." Ibid. 109 "through the Sierra de las Minas mountain range . . ." Joseph R. Mendelson, Megan E. B. Jones, A. P. Pessier, Gabriela Toledo, Edward H. Kabay, and Jonathan A. Campbell, "On the Timing of an Epidemic of Amphibian *Chytridiomycosis* in the Highlands of Guatemala." *South American Journal of Herpetology* 9, no. 2 (August 2014): 151–3. 109 "There is no evidence . . ." Pounds, "Amphibians and Reptiles," op. cit., 161. 110 "There on the highline of Panama's Central Cordillera . . ." Moore, op. cit., 54. 110 "Lips and her team swabbed the frogs . . ." Collins and Crump, op. cit., 152. 110 "We spent six years waiting for something to happen." Lips, "Witnessing extinction in real time," op. cit. 111 "an article in Nature titled 'Clouded Futures.'" Pounds and Puschendorf, op. cit., 427. 111 "a major paper entitled 'Extinction Risk from Climate Change.'" Quotes and details in the following section are drawn from the 2004 *Nature* paper: Chris D. Thomas et al., "Extinction Risk from Climate Change," *Nature* 427 (2004): 145-148. https://doi.org/10.1038/nature02121. 111 "I immediately felt panic . . ." Lips, "Witnessing extinction in real time," op. cit. 112 "The first infected frog at El Copé . . ." Karen R. Lips, et al., "Emerging Infectious Disease and the Loss of Biodiversity in a Neotropical Amphibian Community," *Proceedings of the National Academy of Sciences* 103, no. 9 (2006): 3165–70. https://doi.org/10.1073/pnas.0506889103. 112 "On October 4, the first dead amphibian . . ." Ibid. 112 "The fungal zoospores reached their victim . . ." "Chytrid Fungus," *Amphibian Ark*. (2022, May 16). https://www.amphibianark.org/the-crisis/chytrid-fungus/. 113 "heap of broken images . . ." T. S. Eliot, *The Waste Land: A facsimile and transcript of the original drafts including the annotations of Ezra Pound*, ed. Valerie Eliot (Faber and Faber Limited, 2010), 135. 113 "corpses finally began to lessen . . ." Lips, et al., "Emerging Infectious Disease and the Loss of Biodiversity in a Neotropical Amphibian Community," op. cit., 3165–70. 113 *"Batrachochytrium dendrobatidis* was spreading in a wave across Central and South America . . ." Karen R. Lips, Jay Diffendorfer, Joseph R. Mendelson III, and Michael W. Sears, "Riding the Wave: Reconciling the Roles of Disease and Climate Change in Amphibian Declines," *PLoS Biol* 6, no. 3 (2008): 441-454. doi:10.1371/journal.pbio.0060072. 113 "no evidence to support the

ENDNOTES

hypothesis that climate change has been driving outbreaks . . ." Ibid., 453. **114** "**the amphibian chytrid's invasion of naïve populations**" was "the best explanation for the enigmatic, worldwide declines." Collins and Crump, op. cit., 172–3. **114** "**To Alan Pounds, this was an over-simplification** . . ." Pounds and Masters argued that their climate-related hypotheses did not assume that the fungus was native to the regions experiencing declines; accordingly, climate could influence the mechanisms of a disease regardless of where the pathogen came from—whether it was native or invasive. J. Alan Pounds and Karen Masters, "Amphibian Mystery Misread," *Nature* 462 (2009): 38–9. https://doi.org/10.1038/462038a. **115** "**they included a map of Central America** . . ." Lips, et al., "Riding the Wave: Reconciling the Roles of Disease and Climate Change in Amphibian Declines," op. cit., 445. **116** "**Although it is widely assumed** . . ." Ibid., 448. **116** "**Lips had examined museum specimens** . . ." Ibid., 452. **116** "**A more recent study published in 2013** . . ." Kathryn L. Richards-Hrdlicka, "Preserved Specimens of the Extinct Golden Toad of Monteverde (*Cranopsis periglenes*) Tested Negative for the Amphibian Chytrid Fungus (*Batrachochytrium dendrobatidis*)." *Journal of Herpetology* 47, no. 3 (September 2013): 456–8. doi: https://doi.org/10.1670/11-243. **116** "**Bolaños had boots on the ground** . . ." Federico Bolaños, Douglas C. Robinson, and David B. Wake, "A New Species of Salamander (Genus *Bolitoglossa*) from Costa Rica," *Revista De Biologia Tropical* 35 (1987): 87–92.

CHAPTER SIX: INTO BRILLANTE

Unless otherwise noted, details in this chapter were derived from personal communication with the following sources:

Eladio Cruz, 2018 Priscilla Palavicini, 2018
Frank Hensley, 2023 Jay Savage, 2019
Karen Masters, 2016

130 "**Wolf Guindon had been on the hunt** . . ." Crump, op. cit., 162–3. **130** "**But he'd found the pools empty** . . ." Ibid. **130** "**For the second year in a row** . . ." Ibid., 163. **131** "**One more trip—una más** . . ." Chornook and Guindon, op. cit., 69. **131** "**he scribbled off a letter to Marty Crump** . . ." Crump, op. cit., 191. **131** "**he and Giovanni Bello had received funding** . . ." Pounds, "Disappearing Gold," op. cit., 814. Phillips, op. cit., 140. **131** "**they walked the windswept trail** . . ." Pounds et al., "Tests of Null Models for Amphibian Declines on a Tropical Mountain," op. cit., 1307–22. **131** "**brimming with water** . . ." Pounds, "Disappearing Gold," op. cit., 813. **132** "**the hypothesis that they were alive and breeding in other areas** . . ." Ibid., 817. **132** "**After more than a decade of no sightings** . . ." "Golden Toad," The IUCN Red List of Threatened Species. https://www.iucnredlist.org/species/3172/54357699. **132** "**one of the first amphibians in the world to receive the designation.**" Leenders, op. cit., 180. **132** "**a team of biologists from the United States and Mexico** . . ." "Rediscover the Golden Toad," Kickstarter, last updated February 7, 2014. https://www.kickstarter.com/projects/739230775/rediscover-the-golden-toad?ref=discovery_location. **133** "**It's still out there. I am convinced** . . ." "The Search for Costa Rica's Extinct Golden Toad." The Tico Times, 2014: https://ticotimes.net/2014/01/15/the-search

ENDNOTES

-is-on-for-costa-ricas-extinct-golden-toad. **133** "**$22 of proposed costs...**" "Rediscover the Golden Toad," Kickstarter. **139** "**the San Luis waterfall, another haunted place...**" Years earlier, a local biologist had perished in a landslide at the waterfall, after saving the lives of a group of students. **142** "**a damp gust, bringing rain.**" Eliot, op. cit., 145.

CHAPTER SEVEN: THE SURVIVORS

Unless otherwise noted, details in this chapter were derived from personal communication with the following sources:

>   Giovanni Bello, 2019 and 2024   Priscilla Palavicini, 2024
>   Martha Crump, 2018   Alan Pounds, 2018
>   Eladio Cruz, 2018   George Powell, 2024
>   Frank Hensley, 2023   Lindsay Stallcup, 2018
>   Mynor Káshö, 2017 and 2021
>   Karen Masters, 2020   Mills Tandy, 2023
>   Robin Moore, 2019   Mark Wainwright, 2021
>   Ulises Morales, 2017 and 2021   Justin Yeager, 2024

145 "**there is a little-used trail...**" Details and quotations from the rediscovery of the green-eyed frog are drawn from personal conversations between Mark Wainwright and the authors in June 2021. Additional details come from a project report by Mark Wainwright and Andrew Gray titled "Conservation of the Critically Endangered Green-eyed Frog: *Lithobates vibicarius.*" https://portal-cct.com/blobs/cct/30/2021/4/2008_Conservation_Lithobates_vibicarius_Wainwright_Gray.pdf. **145** "**Mark Wainwright was hiking along those steep trails...**" Leenders, op. cit., 482. **146** "**the once-abundant green-eyed frogs (Lithobates vibicarius) had disappeared...**" Pounds, et al., "Tests of Null Models for Amphibian Declines on a Tropical Mountain," op. cit., 1307–22. Leenders, op. cit., 482. **146** "**a pair of young biology students named Justin Yeager and Mark Pepper...**" Details of the *Atelopus varius* rediscovery are derived primarily from a personal conversation with Justin Yeager; the following sources were also consulted: "Young UD Scientist Finds Frog Thought to Be Extinct," *University of Delaware Archive*, 2004 (from "Horizons," College of Agriculture and Natural Resources): https://www1.udel.edu/PR/UDaily/2005/jul/frog070204.html. Moore, op. cit., 81. **147** "**more than a hundred populations of Atelopus varius...**" Leenders, op. cit., 157–8. Mason Ryan, Erick Berlin, and Ron Gagliardo, "Further Exploration in Search of *Atelopus varius* in Costa Rica," *Froglog* 69 (2005): 1–2. Cesar Barrio-Amorós and Juan Abarca, "Another Surviving Population of the Critically Endangered *Atelopus varius* (Anura: *Bufonidae*) in Costa Rica," *Mesoamerican Herpetology* 3, no. 1 (2016): 128–34. **147** "**the Río Lagarto harlequin toads were in fact an entirely different species...**" César Barrio-Amorós, Gerardo Chaves, and Robert Puschendorf, "Current Status and Natural History of the Critically Endangered Variable Harlequin Toad (*Atelopus varius*) in Costa Rica," *Reptiles & Amphibians* 28, no. 3 (December 2021): 374. https://www.crwild.com/_files/ugd/ea5bc6_fb5c805124de4c1ab0c80909f45dde7a.pdf?index=true. Juan P. Ramírez, César A. Jaramillo, Erik D. Lindquist, Andrew J.

ENDNOTES

Crawford, and Roberto Ibáñez, "Recent and Rapid Radiation of the Highly Endangered Harlequin Frogs (*Atelopus*) into Central America Inferred from Mitochondrial DNA Sequences." *Diversity* 12, no. 9 (2020): 1–21. https://doi.org/10.3390/d12090360. **147 "Bolaños's DNA analysis had confirmed..."** "Young UD Scientist Finds Frog Thought to Be Extinct," op. cit. **147 "a student named Juan Abarca was on a field trip..."** Details and quotations from the rediscovery of *Bufo holdridgei* are drawn from a personal conversation between Juan Abarca, Trevor Ritland, and Priscilla Palavicini in June 2024. Additional details are drawn from: Juan Abarca, Gerardo Chaves, Adrián García-Rodríguez, and Rodolfo Vargas, "Reconsidering Extinction: Rediscovery of *Incilius holdridgei* (Anura: Bufonidae) in Costa Rica After 25 Years." *Herpetological Review* 41, no. 2 (2010): 150–2. **147 "eight tiny toads in the leaf litter..."** Ibid., 151. **147 "Only two species of bufonids..."** Ibid. **147 "The latter had disappeared from the area..."** Ibid. **148 "Abarca returned with a small team..."** Ibid. **148 "(documented as UCR 20671)..."** Ibid. **148 "Over the next few days, they found more..."** Ibid. **148 "Ed Taylor had encountered a single specimen..."** Richard M. Novak and Douglas C. Robinson, "Observations on the Reproduction and Ecology of the Tropical Montane Toad, *Bufo holdridgei* Taylor in Costa Rica," *Revista de Biologia Tropical* 23, no. 2 (1975): 213–37. **148 "the toads were probably fossorial..."** Ibid., 221. **149 "2,765 males had been recorded..."** Leenders, op. cit., 174. **149 "roving balls of clambering toads..."** Novak and Robinson, op. cit., 217. **149 "the closest living relative to the golden toad."** Moore, op. cit., 82. **149 "Juan Abarca and his team once more..."** Abarca et al., op. cit., 151. **149 "In the Indigenous communities of Yorkin and Boruca..."** Details of Bribri amphibian mythology are drawn from conversations with Mynor Káshö in 2017 and 2021. Details of Boruca mythology are drawn from conversations with Ulises Morales in 2017 and 2021. Additional information on amphibian mythologies from other cultures is drawn from Crump's *Eye of Newt...* (2015) and "Emisarias de la Lluvia," a companion booklet to Costa Rica's "Ambassadors of the Rain" exhibition at the Fundación Museos Banco Central de Costa Rica by Priscilla Molina Muñoz and Jennifer L. Stynoski from 2020; citing Corrales (1975), p. 41), the authors write that "indigenous people consider the frog a water deity that brings forth rain and is linked to abundance." Martha L. Crump, *Eye of Newt and Toe of Frog, Adder's Fork and Lizard's Leg* (University of Chicago Press, 2015), 70–2, 84–7. Priscilla Molina Muñós and Jennifer L. Stynoski, Emisarias de la Lluvia: *Anuros en la Época precolombina* (Fundación Museos Banco Central de Costa Rica, 2019), 20–162. **149 "toads appearing from beneath the earth..."** Beau D. Reilly, David I. Schlipalius, Rebecca L. Cramp, Paul R. Ebert, and Craig E. Franklin, "Frogs and Estivation: Transcriptional Insights into Metabolism and Cell Survival in a Natural Model of Extended Muscle Disuse," Physiol Genomics 45, no. 10 (2013), 377–88. DOI: 10.1152/physiolgenomics.00163.2012. **151 "a conservation biologist named Robin Moore."** Moore, op. cit., 86–91. **152 "Moore's own journey on the trail of resurrection..."** Ibid. **152 "Robert Puschendorf had recently rediscovered..."** Moore, op. cit., 82–3. **152 "the first yellow-spotted bell frog *(Litoria castanea)*..."** Ibid., 83. **153 "the challenge of inspiring public engagement..."** Ibid., 86–91. **153 "'If there are survivors,' Moore wondered..."** Robin Moore, personal communication, 2019. 153 "Moore and his colleagues put out a call . . ." Ibid., 86. 153 "some species had been documented . . ." Ibid., 89–90. 153 "McNeil replied 'I fucking love it.'" Ibid., 88. 154 "they coordinated with local scientists . . ." Ibid. 154 "The idea behind the top ten . . ." Ibid., 89. 154 "The gastric-brooding frog found a place . . ." Ibid., 89–91. 154 "Conservation International announced the campaign . . ." Ibid., 86–7. 155 "Moore spent the early days of the Search for Lost Frogs . . ." Ibid., 97. 155 "the Australian team was using mapping technologies . . ." Ibid., 93. 155 "in Borneo, local development . . ." Ibid., 95. 155 "the local teams reported tropical storms . . ." Ibid., 94–5. 155 "the Revolutionary Armed Forces of Colombia . . ." Ibid., 100–4. 155 "All of Colombia's lost frogs . . ." Ibid., 127. 155 "Biologists in Africa's Ivory Coast . . ." Ibid., 130–1. 155 "rediscovery of the cave splayfoot salamander. . ." Ibid., 131–3. 156 "a stream-side sighting of a black-and-yellow frog. . ." Ibid., 176. 156 "the rediscovery of fifteen lost frogs. . ." Ibid., 178. 156 "the Borneo team emailed Moore . . ." Ibid., 179. Robin Moore, "Race for the Rainbow Toad," *Re:wild*, 2016. https://www.rewild.org/news/race-for-the-rainbow-toad. 156 "And in November of that year . . ." Moore, op. cit., 189. 156 "the 'Search for Lost Species' . . ." "Search for Lost Species," *Global Wildlife Conservation (now Re:wild)*. https://www.globalwildlife.org/search-for-lost-species/. 156 "the Jackson's climbing salamander in Guatemala . . ." "Jackson's Climbing Salamander," *Re:wild*, 2019. https://www.rewild.org/lost-species/jacksons-climbing-salamander. Shreya Dasgupta, "Brilliantly colored 'lost' salamander rediscovered after 42 years," *Mongabay*, 2017: https://news.mongabay.com/2017/11/brilliantly-colored-lost-salamander-rediscovered-after-42-years/. Jason Bittel, "'Golden Wonder' Rediscovered After 42 Years," *National Geographic,* 2017: https://www.nationalgeographic.com/animals/article/salamanders-extinct-species-rediscovery-guatemala. 157 "finally located a mate for 'Romeo'. . ." Lindsay Renick Mayer, Press Release: "Lonely No More: Romeo the Sehuencas Water Frog Finds Love," *Re:wild*, 2019: https://www.rewild.org/press/lonely-no-more-romeo-the-sehuencas-water-frog-finds-love. 157 "we have what appears to be the Fernandina Giant Tortoise . . ." "Fernandina Galápagos Tortoise," *Re:wild*, https://www.rewild.org/lost-species/fernandina-galapagos-tortoise. 157 "the first was the Mindo harlequin toad . . ." Molly Bergen, "Researchers Serendipitously Rediscover Ecuador's Lost Mindo Harlequin Toad," *Re:wild*, 2020: https://www.rewild.org/news/mindo-harlequin-toad-rediscovered-after-30-years. 157 "The second was the starry night harlequin toad . . ." Lindsay Renick Mayer, Press Release: "FOUND: Lost Starry Night Harlequin Toad Makes Radiant Return to Science," *Re:wild*, 2019: https://www.rewild.org/press/found-lost-starry-night-harlequin-toad-makes-radiant-return-to-science. "A Bright Future for the Starry Night Harlequin Toad," *Re:wild*, 2019: https://www.rewild.org/news/a-bright-future-for-the-starry-night-harlequin-toad. 157 "Four years of dialogue between the Colombian NGO Fundación Atelopus . . ." "Building Bridges for the Sierra Nevada de Santa Marta," *Re:wild*, 2019: https://www.rewild.org/news/building-bridges-for-the-sierra-nevada-de-santa-marta. 158 "they released a list of ten 'Lost Legends.'" "Lost Legends," *Re:wild*, https://www.rewild.org/lost-species/lost-legends. Milo Putnam, "These are the world's lost and lingering legends," *Re:wild*, 2022: https://www.rewild.org/news

ENDNOTES

/these-are-the-worlds-lost-and-lingering-legends. 159 **"long shots at best."** "Lost Legends," Re:wild. 159 **"An electrical storm came in . . ."** Wainwright and Gray, op. cit. 159 **"they had slept in the old cabin . . ."** Ibid. 159 **"they'd found several hundred green-eyed frogs . . ."** Ibid. 160 **"Ensuring the long-term survival . . ."** Ibid. 160 **"the group included park guards . . ."** Ibid. 160 **"a two-day amphibian workshop . . ."** Ibid. 160 **"the forest guards were still stealing away . . ."** Ibid. 161 **"The local team began monitoring . . ."** Abarca et al., op. cit., 151–2. 161 **"artificial pools to encourage reproduction."** "Conservation of Holdridge's Toad," *Costa Rica Wildlife Foundation*. https://costaricawildlife.org/conservation-of-holdridge-toad/. 161 **"a popular (but illegal) tourist destination . . ."** Leenders, op. cit., 175. 161 **"the toads had only been found at two sites . . ."** Ibid., 174. 161 **"Surveys beginning in 2005 . . ."** Ryan, Berlin, and Gagliardo, op. cit., 1–2. Barrio-Amorós and Abarca, op. cit., 128–34. 161 **"no more than thirty-one individual toads . . ."** Barrio-Amorós, Chaves, and Puschendorf, op. cit., 382. https://www.crwild.com/_files/ugd/ea5bc6_fb5c805124de4c1ab0c80909f45dde7a.pdf?index=true. 161 **"a second population was discovered in the Las Tablas Protected Zone . . ."** Barrio-Amorós, Chaves, and Puschendorf, op. cit., 374–88. José F. González-Maya, et al., "Renewing Hope: The Rediscovery of Atelopus varius in Costa Rica." *Amphibia-Reptilia* 34 (2013): 573–8. DOI:10.1163/15685381-00002910. 161 **"One individual was later reported near Buenos Aires . . ."** Barrio-Amorós and Abarca, op. cit., 128–34. J. C. Solano-Cascante, B. J. Solano-Cascante, E. E. Boza-Oviedo, J. Vargas-Quesada, and D. Sandí-Méndez, "Hallazgo del sapo payaso *Atelopus varius* (Anura: *Bufonidae*) en La Luchita (Potrero Grande: Buenos Aires: Puntarenas: Costa Rica)," *Nota Informativa / Proyecto Biodiversidad de Costa Rica*, San José, Costa Rica (2014). 162 **"and other populations were documented in western Panama."** Barrio-Amorós and Abarca, op. cit., 128–34. Rachel Perez, et al., "Field Surveys in Western Panama Indicate Populations of *Atelopus varius* Frogs Are Persisting in Regions where *Batrachochytrium dendrobatidis* Is Now Enzootic." *Amphibian & Reptile Conservation* 8, no. 2 (2014): 30–5. 162 **"the herpetologists César Barrio-Amorós and Juan Abarca . . ."** Barrio-Amorós and Abarca, op. cit., 128–34. 162 **"Of the 169 localities . . ."** Barrio-Amorós, Chaves, and Puschendorf, op. cit., 374–88. César L. Barrio-Amorós, "The Variable Harlequin Toad in Costa Rica: Raising Hope or Threatened with Extinction?" *Responsible Herpetoculture Journal* (March-April 2024): 64. Barrio-Amorós and Abarca, op. cit., 128–34. González-Maya et al., op. cit., 573–8. 162 **"and deforestation (for palm oil and pineapple and livestock) . . ."** César Luis Barrio-Amorós, "Atelopus, un sueño personal Consideraciones de un herpetólogo neotropical," *Atelopus: Un género en vías de extinción?* (2024): 57–9, 69–71. 162 **"no harlequin frog has been seen in Monteverde . . ."** Leenders, op. cit., 157. 162 **"there are green-eyed frogs in a captive breeding program . . ."** Wainwright and Gray, op. cit.

ENDNOTES

CHAPTER EIGHT: ON THE MOUNTAIN OF REVELATION

Unless otherwise noted, details in this chapter were derived from personal communication with the following sources:

    Gilbert Alvarado, 2021   Luis Solano, 2021
    Eladio Cruz, 2017, 2018, and 2021   Mark Wainwright, 2021
    David Ritland, 2015 and 2021

176 "Whom would the grail serve?" Jessie Weston, From Ritual to Romance (Dover, 1997), 14. 177 "had rediscovered the endemic red-bellied streamside frog . . ." Randall Jiménez and Gilbert Alvarado, "*Craugastor escoces* (Anura: *Craugastoridae*) Reappears After 30 Years: Rediscovery of an 'Extinct' Neotropical Frog," *Amphibia-Reptilia* 38, no. 2 (2017): 257–9. 177 "Gilbert had identified new populations . . ." Andrés Jiménez-Monge, Felipe Montoya-Greenheck, Federico Bolaños, and Gilbert Alvarado, "From Incidental Findings to Systematic Discovery: Locating and Monitoring a New Population of the Endangered Harlequin Toad," *Amphibian and Reptile Conservation* 13, no. 2 (2019): 115–25. 177 "amphibian resistance to *Batrachochytrium dendrobatidis* . . ." Marina E. De León, et al., "*Batrachochytrium dendrobatidis* Infection in Amphibians Predates First Known Epizootic in Costa Rica," *PLoS ONE* 14, no. 12 (2019): https://doi.org/10.1371/journal.pone.0208969. Gilbert Alvarado, J. A. Morales, and Federico Bolaños, "Understanding Patterns of Amphibian Chytrid Fungus Infection and Chytridiomycosis through Museum Collections: Maximizing Sampling and Minimizing Damage on Preserved Specimens," *Herpetological Review* 45, no.1 (2014): 28–31. 181 "Frank Hensley's final observation on Brillante . . ." Philips, op. cit., 20. Moore, op. cit., 26–7. Leenders, op. cit., 180. 186 "I had heard stories of him going out . . ." Chornook, "From the House in the Hole," Walking with Wolf blog: https://walkingwithwolf.wordpress.com/tag/mark-wainwright/. 193 "less than 250 mature individuals of the species . . ." "Narrow-Lined Treefrog," The IUCN Red List of Threatened Species, https://dx.doi.org/10.2305/IUCN.UK.2020-3.RLTS.T55390A54345829.en.

CHAPTER NINE: BURIAL OF THE DEAD

Unless otherwise noted, details in this chapter were derived from personal communication with the following sources:

    Greg Czechura, 2023   Joseph Mendelson, 2024
    Karen Masters, 2024   Lindsay Stallcup, 2018

199 "This frog's name is Toughie . . ." Jason Daly, "Adiós, Toughie: The Last Known Rabb's Fringe-Limbed Tree Frog Dies in Atlanta," Smithsonian Magazine, October 4, 2016. https://www.smithsonianmag.com/smart-news/adios-toughie-last-rabbs-fringe-limbed-tree-frog-dies-atlanta-180960671/. 199 "locked away in a biosecure facility . . ." "Atlanta Botanical Garden frogPOD," *Atlanta Botanical Garden*, https://web.archive.org/web/20161002110714/http://atlantabg.org/learn/conservation-efforts/amphibian-conservation. 199 "a herpetologist named Mark Mandica . . ." Brian Handwerk,

ENDNOTES

"Famous Frog Toughie Dies, Sending Species to Extinction," *National Geographic*, September 30, 2016, https://www.nationalgeographic.com/animals/article/toughie-rabbs-fringe-limbed-tree-frog-dies-goes-extinct. **199 "the adult frogs had been reluctant to mate. . ."** Dante Fenolio, "In Memory of Rabbs' Fringelimb Treefrog," Facebook, September 12, 2016. https://www.facebook.com/dante.fenolio/posts/1554766527882366. **199 "the last known female of the species passed away."** Mike Gaworecki, "Rabbs' Fringe-Limbed Tree Frog Now Presumed to Be Extinct," *Mongabay*, October 7, 2016, https://news.mongabay.com/2016/10/rabbs-fringe-limbed-tree-frog-now-presumed-to-be-extinct/. **199 "In 2012, the health of one of the last two males . . ."** "It's Leap Year. Remember the Rabbs' Tree Frog," *Zoo Atlanta*, February 17, 2012, https://web.archive.org/web/20120524060207/http://www.zooatlanta.org/home/article_content/oleapyear. **200 "news stories and documentary films . . ."** Mark Mandica, "Our Rabbs' Fringed Limbed Tree Frog and Other Species Projected on the UN Building Last Weekend!" *FrogPod Blog*, September 24, 2014, https://web.archive.org/web/20141006202349/http://frogpodblog.blogspot.com/2014/09/our-rabbs-fringed-limbed-tree-frog-and.html. **200 "I will find the male dead of natural causes . . ."** Joseph Mendelson, "Shifted Baselines, Forensic Taxonomy, and Rabbs' Fringe-Limbed Treefrog: The Changing Role of Biologists in an Era of Amphibian Declines and Extinctions." *Herpetological Review* 42, no. 1 (2011): 21–5. **200 "He would have found a mate by calling for . . ."** Daley, op. cit. **200 "in 2014, Toughie called again."** Mark Mandica, "The Rabbs' Fringe-Limbed Tree Frog Called Again . . . After Years . . . and We Recorded It!" *The Amphibian Foundation's Frog Blog*, December 17, 2014, https://amphibianfound.blogspot.com/2014/12/the-rabbs-fringe-limbed-tree-frog.html. **201 "Toughie died sometime in the early morning . . ."** Daley, op. cit. **204 "in situ . . . conservation . . ."** "In situ conservation," *ScienceDirect*, https://www.sciencedirect.com/topics/agricultural-and-biological-sciences/in-situ-conservation. **205 "Thousand Frog Stream . . ."** Elizabeth Kolbert, "The Sixth Extinction?" *The New Yorker*, May 18, 2009. https://www.newyorker.com/magazine/2009/05/25/the-sixth-extinction. **205 "an inventory of the species . . ."** R. Gagliardo, P. Crump, E. Griffith, J. Mendelson, H. Ross, and K. Zippel, "The Principles of Rapid Response for Amphibian Conservation, Using the Programmes in Panama as an Example." *International Zoo Yearbook* 42, no. 1 (2008): 125–35. https://doi.org/10.1111/j.1748-1090.2008.00043.x. **207 "with support from the Houston Zoo . . ."** "History," Amphibian Rescue and Conservation Project, http://amphibianrescue.org/about/history/. **207 "Hundreds of frogs who had missed the departing ark . . ."** Jenni Laidman, "Rescuers Race to Save Central American Frogs." *The Blade*, August 6, 2006. https://www.toledoblade.com/frontpage/2006/08/06/Rescuers-race-to-save-Central-American-frogs.html. **207 "Jay Savage had once written . . ."** Jay Savage, "On the Trail of the Golden Frog: With Warszewicz and Gabb in Central America." *Proceedings of the California Academy of Sciences* 38, no. 4 (1970): 273–87. **207 "chytrid had traversed the Panama Canal . . ."** Douglas C. Woodhams, et al., "Chytridiomycosis and Amphibian Population Declines Continue to Spread Eastward in Panama," *EcoHealth* 5, no. 3 (2008): 268–74. https://doi.org/10.1007/s10393-008

ENDNOTES

-0190-0. 207 "**deflating hopes that the barrier would slow the spread . . .**" Rhett A. Butler, "Armageddon for Amphibians? Frog-Killing Disease Jumps Panama Canal." *Mongabay Environmental News*, October 12, 2008. https://news.mongabay.com/2008/10/armageddon-for-amphibians-frog-killing-disease-jumps-panama-canal/#:~:text=Chytridiomycosis%20%E2%80%94%20a%20fungal%20disease%20that,amphibians%20east%20of%20the%20canal. 208 "**Other biologists in attendance founded the Amphibian Ark . . .**" "Amphibians Need Your Help: Jump In!" *Amphibian Ark*, https://www.amphibianark.org/pdf/AArk-5-panel-brochure.pdf. 208 "**It is not the goal of AArk's programs . . .**" "About Us," *Amphibian Ark*, https://www.amphibianark.org/about-us/. 208 "**'Desperate though they may be,' they wrote . . .**" Joseph R. Mendelson III, et al., "Confronting Amphibian Declines and Extinctions," *Science* 313 (2006): 48. DOI:10.1126/science.1128396. 209 "***Batrachochytrium dendrobatidis* has been responsible . . .**" Ben C. Scheele, et al., "Amphibian Fungal Panzootic Causes Catastrophic and Ongoing Loss of Biodiversity," *Science* 363, no. 6434 (2019): 1459–63. DOI:10.1126/science.aav0379. 209 "**Scientists are still discovering new lineages . . .**" Allison Q. Byrne, et al., "Cryptic Diversity of a Widespread Global Pathogen Reveals Expanded Threats to Amphibian Conservation." *Proceedings of the National Academy of Sciences* 116, no. 41 (2019): 20382–7. https://doi.org/10.1073/pnas.1908289116. 209 "**In 2018, a Brazilian hybrid was found . . .**" S. E. Greenspan, et al., "Hybrids of Amphibian Chytrid Show High Virulence in Native Hosts." *Scientific Reports* 8, no. 1 (2018). https://doi.org/10.1038/s41598-018-27828-w. 209 "**a 2023 study confirmed . . .**" Tamilie Carvalho, et al., "Coinfection with Chytrid Genotypes Drives Divergent Infection Dynamics Reflecting Regional Distribution Patterns." *Communications Biology* 6, 941 (2023). https://doi.org/10.1038/s42003-023-05314-y. 210 "**temperature variability can make it more difficult . . .**" Thomas R. Raffel, John M. Romansic, Neal T. Halstead, Taegan A. McMahon, Matthew D. Venesky, and Jason R. Rohr, "Disease and Thermal Acclimation in a More Variable and Unpredictable Climate." *Nature Climate Change* 3, no. 2 (2012): 146–51. https://doi.org/10.1038/nclimate1659. 210 "**The survivable range of amphibians is shrinking.**" Jennifer A. Luedtke, et al., "Ongoing Declines for the World's Amphibians in the Face of Emerging Threats." *Nature* 622 (2023): 308–14. https://doi.org/10.1038/s41586-023-06578-4. 210 "**mounting research has continued to vindicate . . .**" Stefan Lötters, et al., "Ongoing Harlequin Toad Declines Suggest the Amphibian Extinction Crisis Is Still an Emergency." *Communications Earth & Environment* 4, no. 412 (2023). https://doi.org/10.1038/s43247-023-01069-w. 210 "**one of two 'climate change associated global species extinctions to date.'**" Hans-Otto Pörtner, et al., eds., "Climate Change 2022: Impacts, Adaptation and Vulnerability" (Intergovernmental Panel on Climate Change, 2022). https://www.ipcc.ch/report/ar6/wg2/downloads/report/IPCC_AR6_WGII_FullReport.pdf. 211 "**In June 2022, two Australian scientists . . .**" Jodi Rowley and Karrie Rose, "Australian Frogs Are Dying En Masse Again, and We Need Your Help to Find Out Why." *The Conversation*, September 11, 2023. https://theconversation.com/australian-frogs-are-dying-en-masse-again-and-we-need-your-help-to-find-out-why-184255. 212 "**southern Africa had been the origin . . .**" Ché Weldon, Louis H. du

ENDNOTES

Preez, Alex D. Hyatt, Reinhold Muller, and Rick Speare, "Origin of the Amphibian Chytrid Fungus." *Emerging Infectious Diseases* 10, no. 12 (December 2004): 2100–5. https://doi.org/10.3201/eid1012.030804. 212 **"*Bd* has expanded its foothold in Africa . . ."** Sonia L. Ghose, et al., "Continent-Wide Recent Emergence of a Global Pathogen in African Amphibians." *Frontiers in Conservation Science*, 4 (2023). https://doi.org/10.3389/fcosc.2023.1069490. 212 **"disease-driven declines and extinctions of amphibians may already be occurring in Africa . . ."** "Rapid Surge in Highly Contagious Killer Fungus Poses New Threat to Amphibians across Africa," *Phys.org*, March 15, 2023. https://phys.org/news/2023-03-rapid-surge-highly-contagious-killer.html. 212 **"researchers discovered evidence of chytrid in Madagascar frogs . . ."** Jonathan E. Kolby, "Presence of the Amphibian Chytrid Fungus *Batrachochytrium dendrobatidis* in Native Amphibians Exported from Madagascar." *PLoS ONE* 9, no. 3 (2014). https://doi.org/10.1371/journal.pone.0089660. 212 **"the pathogen had been present on the island . . ."** Molly C. Bletz, et al., "Widespread Presence of the Pathogenic Fungus *Batrachochytrium dendrobatidis* in Wild Amphibian Communities in Madagascar." *Scientific Reports* 5, no. 8633 (2015). https://doi.org/10.1038/srep08633. 212 **"I dread the day . . ."** Lips, "What If There Is No Happy Ending? Science Communication as a Path to Change," op. cit. 213 **"Kolby launched an ambitious plan . . ."** Brian Clark Howard, "How to Rescue These Adorable Tree Frogs." *National Geographic*, April 7, 2016. https://www.nationalgeographic.com/animals/article/160407-frog-rescue-chytrid-fungus-honduras-cusuco-jonathan-kolby#:~:text=American%20biologist%20Jonathan%20Kolby%20hopes,northwestern%20part%20of%20the%20country. 213 **"Kolby organized an Indiegogo campaign . . ."** Jonathan Kolby, "Save Frogs From Extinction!" Indigogo, last updated March 15, 2018. https://www.indiegogo.com/projects/save-frogs-from-extinction#/. 213 **"Beginning in 2009, a Spanish ecologist . . ."** Lizzie Wade, "A Fungus That's Killing Amphibians All Over the World Finally Meets Its Match." *Wired*, November 18, 2015. https://www.wired.com/2015/11/a-frog-killing-fungus-finally-meets-its-match-on-the-island-of-mallorca/. 214 **"In November 2015, the team published their report . . ."** Jaime Bosch, Eva Sanchez-Tomé, Andrés Fernández-Loras, Joan A. Oliver, Matthew C. Fisher, and Trenton W. J. Garner, "Successful Elimination of a Lethal Wildlife Infectious Disease in Nature." *Biology Letters* 11, no. 11 (2015). https://doi.org/10.1098/rsbl.2015.0874. 214 **"The battle against chytrid began gaining ground . . ."** Donald K. Nichols and Elaine W. Lamirande, "Successful Treatment of Chytridiomycosis." *Froglog* 46 (August 2001). https://www.amphibians.org/wp-content/uploads/sites/3/2018/12/Froglog46.pdf. 214 **"Lee Berger's lab in Australia developed an initial detection swab . . ."** Berger, et al., "Chytridiomycosis Causes Amphibian Mortality Associated with Population Declines in the Rain Forests of Australia and Central America," op. cit., 9031–6. 214 **"a 2014 study found that an antifungal drug called Nikkomycin Z . . ."** Whitney M. Holden, J. Scott Fites, Laura K. Reinert, and Louise A. Rollins-Smith, "Nikkomycin Z Is an Effective Inhibitor of the Chytrid Fungus Linked to Global Amphibian Declines." *Fungal Biology* 118, no. 1 (2014): 48–60. https://doi.org/10.1016/j.funbio.2013.11.001. 214 **"In 2021 . . ."** Anthony W. Waddle, et al., "Amphibian Resistance to Chytridiomycosis Increases Following

ENDNOTES

Low-Virulence Chytrid Fungal Infection or Drug-Mediated Clearance." *Journal of Applied Ecology* 58, no. 10 (2021): 2053–64. https://doi.org/10.1111/1365-2664.13974. 214 **"and 2023 . . ."** Samantha A. Siomko, et al., "Selection of an Anti-Pathogen Skin Microbiome Following Prophylaxis Treatment in an Amphibian Model System." *Philosophical Transactions of the Royal Society B: Biological Sciences* 378, no. 1882 (2023). https://doi.org/10.1098/rstb.2022.0126. 215 **"Studies in 2014 . . ."** Taegan A. McMahon, et al., "Amphibians Acquire Resistance to Live and Dead Fungus Overcoming Fungal Immunosuppression." *Nature* 511, no. 7508 (2014): 224–7. https://doi.org/10.1038/nature13491. 215 **"2019 . . ."** Tiffany A. Kosch, et al., "Genetic Potential for Disease Resistance in Critically Endangered Amphibians Decimated by Chytridiomycosis." *Animal Conservation* 22 (2019): 238–50. https://doi.org/10.1111/acv.12459. 215 **"and 2021 . . ."** Juan G. Abarca, et al., "Genotyping and Differential Bacterial Inhibition of *Batrachochytrium dendrobatidis* in Threatened Amphibians in Costa Rica." *Microbiology* 167, no. 3 (2021). https://doi.org/10.1099/mic.0.001017. 215 **"through the use of other locally adapted skin bacteria . . ."** Ibid. 216 **"And I hope we are wrong about the golden toad . . ."** Mendelson, "Shifted Baselines, Forensic Taxonomy, and Rabbs' Fringe-Limbed Treefrog: The Changing Role of Biologists in an Era of Amphibian Declines and Extinctions," op. cit., 24. 216 **"The first smoke signals . . ."** Annemarieke Spitzen-van der Sluijs, et al., "Rapid Enigmatic Decline Drives the Fire Salamander *(Salamandra salamandra)* to the Edge of Extinction in the Netherlands." *Amphibia-Reptilia* 34, no. 2 (2013): 233–9. https://doi.org/10.1163/15685381-00002891. 217 **"losing 57 percent of their range . . ."** Ibid., 233. 217 **"they were found only in a few isolated sections . . ."** Ibid., 236. 217 **"Monitored since 1971, the Bunderbos salamanders . . ."** Ibid., 234. 217 **"either an infectious agent or intoxication."** Ibid., 237. 217 **"surveys turned up fewer and fewer healthy salamanders . . ."** Ibid., 236. 218 **"The decline we describe . . . strongly resembles the population . . ."** Ibid., 237. 218 **"populations had fallen to 4 percent . . ."** An Martel, et al., "*Batrachochytrium salamandrivorans* sp. nov. Causes Lethal Chytridiomycosis in Amphibians," *Proceedings of the National Academy of Sciences* 110, no. 38 (2019): 15325–9. https://doi.org/10.1073/pnas.1307356110. 218 **"As An Martel performed necropsies . . ."** Katie L. Burke, "New Disease Emerges as Threat to Salamanders." *American Scientist* 103, no. 1 (2015): 6. https://www.americanscientist.org/article/new-disease-emerges-as-threat-to-salamanders. 218 **"the infected salamanders under observation . . ."** Martel, et al., op. cit., 15326. 219 **"If the disease continues to progress at the same rate . . ."** Burke, op. cit., 6.

CHAPTER TEN: THE ETERNAL FOREST

Unless otherwise noted, details in this chapter were derived from personal communication with the following sources:

JJ Apodaca, 2019 and 2024
Kyle Barrett, 2019
Giovanni Bello, 2024
Eladio Cruz, 2017, 2018, and 2021

Will Harlan, 2019 and 2024
Mark Mandica, 2019
Joseph Mendelson, 2024
Ben Prater, 2019 and 2024

ENDNOTES

Lindsay Stallcup, 2018
Donald Varela Soto, 2024
Mark Wainwright, 2021

224 "The English colonizer Walter Raleigh..." Daniel Genkins, "'To Seek New Worlds, for Gold, for Praise, for Glory': El Dorado and Empire in Sixteenth-Century Guiana." *The Latin Americanist* 58, no. 1 (2014): 94. **224 "he lost his son..."** Ibid., 93.Genkins cites: Naipaul, *The Loss of El Dorado*, 85–91. Keith Thomson, *Paradise of the Damned* (Little, Brown, 2024), 259–60, 263. **224 "and then his own life..."** Thomson, op. cit., 301. **224 "Antonio de Berrío, had already undertaken three expeditions..."** Genkins, op. cit., 91. **224 "enfermedad doradista..."** Ibid., 90. **224 "Berrío had discovered a clause..."** Ibid., 90–1. **224 "Berrío died at seventy..."** Ibid., 91. **224 "Berrío's son died of the plague..."** Ibid., 95. **227 "Walter Raleigh claimed that he could see the golden towers..."** John Silver, "The Myth of El Dorado." *History Workshop Journal* 34, no. 1 (1992): 2. **227 "Donald Varela Soto heard a strange call..."** Details and quotations from the discovery of the Tapir Valley tree frog are drawn from a personal conversation with Donald Varela Soto in 2024 and the following sources: Donald Varela Soto, Juan G. Abarca, Esteban Brenes-Mora, Valeria Aspinall, Twan Leenders, and Alex Shepack, "A New Species of Brilliant Green Frog of the Genus *Tlalocohyla* (Anura, *Hylidae*) Hiding Between Two Volcanoes of Northern Costa Rica." *Zootaxa* 5178, no. 6 (2022): 501–31. https://www.mapress.com/zt/article/view/zootaxa.5178.6.1. Corryn Wetzel, "New Frog Species Discovered in Tapir Valley Nature Reserve, Bijagua, Costa Rica." *Re:wild*, 2022: https://www.rewild.org/news/new-frog-species-discovered-in-tapir-valley-nature-reserve-bijagua-costa. Press Release: "A New Tiny Green Frog with a Blue Armpit and Red Spots Has Been Discovered in Costa Rican Nature Reserve." *Re:wild*, 2022: https://www.rewild.org/press/a-new-tiny-green-frog-with-a-blue-armpit-and-red-spots-has-been-discovered. Liz Kimbrough, "Tiny New Tree Frog Species Found in Rewilded Costa Rican Nature Reserve." *Mongabay*, 2022: https://news.mongabay.com/2022/09/tiny-new-tree-frog-species-found-in-rewilded-costa-rican-nature-reserve/. **227 "a week of unusually heavy rains..."** Wetzel, op. cit. **228 "It would take six months of searching..."** Ibid. **229 "a biological corridor for the endangered Baird's tapir..."** Varela Soto, et al., op. cit., 501–31. **229 "It was on an evening in October..."** Ibid. **230 "The small frog was bright green..."** Description of the species is drawn from photographs by Marco Molina. **230 "its vocal sac swelled like a balloon..."** Varela Soto, et al., op. cit., 501–31. And photographs by Marco Molina. **230 "It would take more than three years of writing..."** Varela Soto, et al., op. cit., 501–31. **230 "There were fires smoldering on the Pacific slope..."** Leslie J. Burlingame, "History of The Monteverde Conservation League and The Children's Eternal Rainforest," 2019. https://acmcr.org/wp-content/uploads/2024/07/MCL-History-2019-English-jul2024.pdf. Chornook and Guindon, op. cit., 113. **231 "a homesteading family was busy taking over abandoned properties..."** Chornook and Guindon, op. cit., 114. **231 "Alan Pounds had convened a group of locals..."** Richard LaVal, "Forming of the Monteverde Conservation League," in *Monteverde Jubilee Family Album*, edited by Lucille Guindon, et al. (Associacion De Amigos De Monteverde, 2001), 203. **231 "Among the biologists and nature lovers..."** LaVal, op. cit., 203. Chornook and Guindon, op. cit., 113. **231 "the group**

ENDNOTES

continued to convene . . ." LaVal, op. cit., 203. 232 "formally founded the Monteverde Conservation League." Ibid. Chornook and Guindon, op. cit., 114. 232 "the League had turned its attention largely to the Peñas Blancas valley . . ." Burlingame, "History of The Monteverde Conservation League and The Children's Eternal Rainforest," op. cit. 232 "Bob Law, Wolf Guindon, and Giovanni Bello . . ." Chornook and Guindon, op. cit., 119. 232 "some landowners were willing . . ." Burlingame, "History of The Monteverde Conservation League and The Children's Eternal Rainforest," op. cit. Ibid., 120. 232 "Eladio Cruz had come over the Continental Divide . . ." As George Powell advised us, it is worth noting that bananas grown for food have no seeds: they are grown from suckers (hios) on the mother plant; Eladio would have been carrying hios. Eladio Cruz, "Libro Vivo" speaker series, 2017. George Powell, personal communication, 2024. 232 "Wolf Guindon formalized the purchase." Chornook and Guindon, op. cit., 115. 232 "dozens of other settlers in the valley lined up . . ." Burlingame, "Conservation in the Monteverde Zone: Contribution of Conservation Organizations," op. cit., 363. 233 "the original funds that the League had obtained . . ." Ibid., 362–3. 233 "a biologist named Sharon Kinsman . . ." Ibid. 233 "Kinsman had been living in Monteverde . . ." K. Greg Murray, Peter Feinsinger, William H. Busby, Yan B. Linhart, James H. Beach, and Sharon Kinsman, "Evaluation of Character Displacement Among Plants in Two Tropical Pollination Guilds," *Ecology* 68, no. 5 (1987): 1283–93. https://doi.org/10.2307/1939213. 233 "The Swedish government agreed to match the funds . . ." Burlingame, "Conservation in the Monteverde Zone: Contribution of Conservation Organizations," op. cit., 363. 233 "to purchase six hectares of forest . . ." Ibid. 233 "The campaign begun by the Swedish students snowballed . . ." Ibid., 364. Chornook and Guindon, op. cit., 117–8. 233 "fundraising efforts spread . . ." Burlingame, "Conservation in the Monteverde Zone: Contribution of Conservation Organizations," op. cit., 364. 233 "the 'Save the Rainforest' initiative . . ." Chornook and Guindon, op. cit., 118. Ibid. 233 "'Children's Tropical Forests UK' . . ." Chornook and Guindon, op. cit., 118. "About Us," *Children's Tropical Forests*, https://www.tropical-forests.com/. 233 "Chico Friends in Unity with Nature in California, Kinderregenwald in Germany, Nippon Kodomo no Jungle in Japan." Burlingame, "Conservation in the Monteverde Zone: Contribution of Conservation Organizations," op. cit., 364. 234 "the new reserve was named 'Bosque Eterno de los Niños' . . ." Ibid., 363. 234 "the Children's Eternal Rainforest had grown . . ." Ibid., 364. "Quienes Somos," *Bosque Eterno de los Niños*, https://acmcr.org/quienes-somos/. 234 "The streams feed rivers . . ." "Mapas," *Bosque Eterno de los Niños*, https://acmcr.org/quienes-somos/mapas/. 234 "the largest private forest . . ." Burlingame, "Conservation in the Monteverde Zone: Contribution of Conservation Organizations," op. cit., 364. 234 "the Monteverde Conservation League embraced this philosophy." Ibid., 364–6. 234 "They worked with local teachers . . ." Ibid. Guillermo Vargas, "The Community Process of Environmental Education," in *Monteverde: Ecology and Conservation of a Tropical Cloud Forest,* edited by Nalini M. Nadkarni and Nathaniel T. Wheelwright (New York: Oxford Academic, 2000), 377–8. 234 "Collaboration with local farmers and landowners . . ." Burlingame, "Conservation in the Monteverde Zone:

ENDNOTES

Contribution of Conservation Organizations," op. cit., 365. **235** "a local climate-change organization called Corclima..." "Corclima," *Corclima*. https://corclima.org/. **235** "The 'Mi Ocotea' project..." "An Exceptional Tree and a Symbol of Monteverde," Monteverde Institute Blog, 2016. https://monteverde-institute-blog.org/environmental/ocotea-monteverdensis. Mi_Ocotea, Instagram. https://www.instagram.com/mi_ocotea. **235** "The Santa Elena Reserve..." Burlingame, "Conservation in the Monteverde Zone: Contribution of Conservation Organizations," op. cit., 367–9. **236** "more than fifty individuals sighted in a single night..." Varela Soto, et al., op. cit., 501–31. **236** "Tlaloc Conservation..." Tlaloc Conservation, Instagram. https://www.instagram.com/tlalocconservation/. **238** "a fragment shored against its ruins..." Eliot, op. cit., 146. **239** "the number one hot spot for salamander biodiversity..." "Amphibians of the World," *Biodiversity Mapping*. https://biodiversitymapping.org/index.php/amphibians/. **240** "There are more species of salamanders there..." Ibid. **242** "still carrying the weight of Toughie's loss..." Ed Yong, "The Last of Its Kind," *The Atlantic* (August 2019). https://www.theatlantic.com/magazine/archive/2019/07/extinction-endling-care/590617/. **242** "the frosted flatwoods salamander..." "Stay Frosty," *The Amphibian Foundation,* https://www.amphibianfoundation.org/index.php/stay-frosty. **242** "it bred in captivity for the first time." Dan Chapman, "Survival of the *'frosties',*" *U.S. Fish and Wildlife*, May 16, 2023. https://www.fws.gov/story/2023-05/survival-frosties. **244** "a partial ban on the import of salamander species..." "U.S. Fish and Wildlife Service Bans Import of Salamander Species in the Face of Deadly Disease," *Defenders of Wildlife*, January 12, 2016. https://defenders.org/newsroom/us-fish-and-wildlife-service-bans-import-of-salamander-species-face-of-deadly-disease. **245** "practices like cloning and resurrection..." Ed Yong, "Resurrecting the Extinct Frog with a Stomach for a Womb," National Geographic, March 15, 2023. https://www.nationalgeographic.com/science/article/resurrecting-the-extinct-frog-with-a-stomach-for-a-womb. **246** "what the thunder said..." Eliot, op. cit., 143.

EPILOGUE: EL DORADO Y LOS ÁNGELES

Unless otherwise noted, details in this chapter were derived from personal communication with the following sources:

    Neftali Camacho, 2021 and 2024   Eladio Cruz, 2021

249 "He was the first specimen..." Savage, "An Extraordinary New Toad (*Bufo*) from Costa Rica," op. cit., 153–67. Excerpt from Norman Scott's field notes from May 14, 1964, shared with the authors. **250** "If I found it within the borders of the Children's Eternal Rainforest..." Eladio Cruz, personal communication, 2021. **252** "a damp gust, bringing rain." Eliot, op. cit., 145.